Energy and Empire

Energy and Empire

The Politics of Nuclear and Solar Power
in the United States

GEORGE A. GONZALEZ

Cover images of nuclear plant and windmills/solar panels courtesy of Fotolia.

Published by State University of New York Press, Albany

For information, contact State University of New York Press, Albany, NY
www.sunypress.edu

Production by Diane Ganeles
Marketing by Fran Keneston

Library of Congress Cataloging-in-Publication Data

Gonzalez, George A., 1969–
   Energy and empire : the politics of nuclear and solar power in the United States / George A. Gonzalez.
       p. cm.
   Includes bibliographical references and index.
   ISBN 978-1-4384-4295-2 (hbk. : alk. paper)
   1. Energy policy—United States.   2. Environmental policy—United States.
3. Nuclear energy—Government policy—United States.   4. Solar energy—
Government policy—United States.   I. Title.

   HD9502.A2G655 2012
   333.792'30973—dc23   2011035851

10 9 8 7 6 5 4 3 2 1

For Ileana and Alana

Contents

Chapter 1

The Politics of Nuclear and Solar Energy

In *Urban Sprawl, Global Warming, and the Empire of Capital*, I explain how and why urban sprawl arose as a lead strategy/means to stabilize the U.S. economy in the 1930s, and later as a lynchpin for the world economy in the post–World War II period, and remains as such.[1] In addition to growing the world capitalist economy, urban sprawl also greatly pushes up energy demand because it creates an energy-intense transportation infrastructure (i.e., automobile dependency) and an energy-intense housing stock (low-density urban development expands energy use to heat/cool and power the appliances that fill the relatively large multiroom households that are characteristic of such development). Whereas the thrust of *Urban Sprawl, Global Warming, and the Empire of Capital* concentrates on why the United States consumes so much energy, in this book I emphasize the supply side of the U.S. energy equation.

The United States, of course, meets its massive energy demand mostly through fossil fuels, which in turn leads to massive greenhouse emissions (roughly 20 to 25 percent of the world's total anthropogenic climate change gasses).[2] There is another important component to U.S. energy consumption: nuclear power (with about 20 percent of its electricity coming from this source).[3] The U.S. development and political sponsorship of nuclear power is the focus of this book. During the 1950s, the U.S. government made the strategic decision to support nuclear power as the energy of the future, neglecting the promise of solar energy (i.e., passive solar, photovoltaic, wind, and wave power).[4] The United States pursued nuclear power in spite of its obvious public health and environmental dangers, including the

1

perils of nuclear weapons proliferation. Part of the irony of the U.S. government's pursuit of nuclear power and virtual ignoring of solar energy[5] is that among advanced industrialized countries, the United States has arguably the greatest ostensive solar potential. I am specifically pointing to its sun- and heat-drenched Southwestern desert region.[6] (The United States also has a very windy Midwest.[7]) Thus, it is possible to conceptualize a scenario in which the United States becomes a major producer of surplus energy by exploiting the sunlight and heat of this desert region.[8]

The United States' nuclear path was set by economic elites (through the Rockefeller Foundation and the Panel on the Impact of the Peaceful Uses of Atomic Energy). Conversely, U.S. economic elites went on the record in the 1950s (via the Association for Applied Solar Energy) in opposing government support for solar power. The result is that nuclear power presently appears as the only viable alternative to fossil fuels, with solar energy an evolving substitute at best.[9] Nuclear power nonetheless is so plagued by economic, safety (e.g., the Fukushima Daiichi disaster), and geopolitical liabilities that in spite of depleting fossil fuels and the obvious dangers of global warming, the world, including the United States, has only moved relatively slowly toward nuclear power as an alternative to what are disappearing and perilous fossil fuels.[10]

More than fifty years after economic elites in the desert Southwest expressed their disapproval of public subsidies for solar power, we are left wondering how far solar power could have been developed if the U.S. government had pursued this form of energy as aggressively as it did nuclear power. It is precisely because there is no economic and safe alternative to fossil fuels[11] that the international community cannot agree to a global strategy to avoid cataclysmic climate change.[12]

The Limits of Nuclear Power

The decision in the 1950s to back nuclear power and not do the same for solar has seemingly profound implications for the present period and for humanity and the environment generally. Safety has remained a key concern with the operation of nuclear power plants[13] (e.g., the recent nuclear meltdowns in Japan[14]), and as a result, the costs of such plants has skyrocketed—as nuclear power plants have to be built/engineered with redundant and sophisticated safety systems. The costs of nuclear power plants have increased over time as larger and larger plants are constructed to achieve higher economies of scale in an effort to push down the costs per unit of energy delivered.[15] The result is that civilian nuclear power can only proceed with heavy government subsidies, including capping firms' liability in the case of a public health disaster arising from the accidental release of radioactive energy (e.g., the Price-Anderson Act).[16] A recently authorized nuclear power plant in the United States has been suspended because the government demanded a greater financial contribution from the operators of the plant.[17]

Another liability of civilian nuclear power relates to politics. Particularly after the Three Mile Island (1979) accident and the one at Chernobyl (1986),[18] much of the public has become leery of nuclear power.[19] (The public apprehension of this source of power is being amplified by the Japanese Fukushima Daiichi nuclear disaster of March 2011.[20]) Local business interests, fearful for the local economic climate,[21] successfully prevented the start-up of a completed nuclear power plant on Long Island, New York, in the 1980s.[22] Potentially decisive local opposition to a finished plant drives up the risk and uncertainties of nuclear power.[23] The Vermont state legislature voted in 2010 not to extend the operating license of the Yankee nuclear power plant.[24] Current plans have the New Jersey Oyster Creek nuclear power plant shutting down in 2019, ten years earlier than initially planned, because of management's unwillingness to comply with the state government's demand for upgrading the safety measures of the plant.[25]

Perhaps the most long-term liability of nuclear power is its waste by-product. Nuclear waste has a half-life that is in the tens of thousands of years. Thus, even if a relatively modest amount of this waste were to contaminate an aquifer, lake, river, or watershed, the water would be unsuitable for consumption for ostensibly eons. There is no storage technology/method currently available that can safely and assuredly store nuclear waste for the entirety of its radioactive life.[26] Hence, a future dominated by nuclear energy would entail ever more waste that poses a grave, intractable threat to human health and the ecosystem for countless generations.[27]

Another liability of nuclear waste arises from the fact that such waste can be mined (processed) for weapons-grade material. There is a distinction between what are known as breeder and light water nuclear reactors.[28] The breeder variant results in more weapons material, but light water reactor waste can also be used to manufacture nuclear weapons.[29] The result is that U.S. foreign policy has been virtually schizophrenic on the matter of other countries developing a domestic nuclear power capacity. In the 1950s, through its Atoms for Peace program, the United States internationally promoted civilian nuclear power.[30] In the 1970s, the Carter administration (1977–1981), because of weapons proliferation concerns, made it a political priority to limit the global trade in civilian nuclear technology.[31]

The Bush and Obama administrations' stance on Indian and Iranian nuclear civilian power programs is particularly contradictory, if not perplexing. India (in an arms race with its neighbor Pakistan) has not signed the Nuclear Non-Proliferation Treaty and has developed nuclear weapons. In spite of this, the U.S. government, under the presidency of George W. Bush, sponsored India's entrance into the international system of civilian nuclear power.[32]

By contrast, Iran is a signatory to the Nuclear Non-Proliferation Treaty and, even according to the United States' intelligence agencies, is in compliance with this treaty (i.e., it is not pursuing a nuclear weapons program).[33] Under the terms of the nonproliferation treaty, because it is not engaged in a weapons program, Iran is within its rights to gain a civilian nuclear capacity.[34] Nevertheless, the U.S. government is leading an effort to block Iran's development of this capacity, even threatening it with military attack to prevent Iran from employing civilian nuclear energy.[35]

Notwithstanding its selective diplomacy on nuclear weapons and energy, it was the United States that opened the door to nuclear weapons and energy. As already noted, the United States promoted nuclear energy beginning in the 1950s. Moreover, the United States has the most civilian nuclear reactors in the world, 104 of a global total of 440, and it is currently building more.[36] Perhaps more importantly, the United States has a huge number of nuclear weapons stockpiled (both limited-range tactical nuclear missiles and intercontinental ballistic nuclear missiles).[37] What makes U.S. diplomacy on nuclear energy particularly unviable and self-defeating is that the United States cannot offer an alternative to fossil fuels and nuclear power. Thus, as fossil fuels deplete in a country like Iran,[38] it is left with one of two alternatives: develop nuclear power in an effort to attain energy independence or become dependent on an international fossil fuel market that is dominated by fewer and fewer producing countries and will be increasingly volatile as more and more of these finite resources are consumed.[39]

Why Nuclear Power?

Historically, why did the United States decide to aggressively sponsor/subsidize the development, deployment, and operation of civilian nuclear power? Why did it not do the same for solar? U.S. economic growth projections in the 1950s and beyond did seemingly create a bias for nuclear power. The full potential of solar remains unknown, but in the short term solar power cannot be adapted to the energy needs created by the U.S. economy, especially when that economy is being spurred by energy-profligate urban sprawl.

But near–medium-term energy demand projections did not, however, prompt the United States to pursue and promote nuclear power. Into the 1950s, the United States was a leading producer of petroleum,[40] and it remains the country with the largest reserves of coal.[41] It also contains massive amounts of natural gas.[42] Moreover, when it came into relief in 1973 that the United States could not meet its domestic oil demand through domestic production, it was able to turn to foreign supplies to meet its needs.[43] Therefore, among advanced industrialized countries in the post–World War II period, the United States was in the best position to forego pursuing civil nuclear power and instead invest in the long-term energy strategy of developing solar energy.

Frank N. Laird, in his book *Solar Energy, Technology Policy, and Institutional Values*, holds that the United States aggressively deployed civilian nuclear power and forewent solar energy because of the ideas that dominated White House thinking on energy. Laird explains that during the post–World War II period, the ideas of energy supply and national security were linked, and beginning with the Eisenhower administration (1953–1961), these concepts became tied to nuclear power—which held the promise of virtually limitless and inexpensive supplies of energy.[44]

Laird, however, fails to grapple with the fact that throughout the postwar period and into the contemporary era, the thinking on energy supply that dominates the U.S. polity is rather unique. This uniqueness results from the position America holds in the global economic/political system. It is the prime leader of the capitalist world system and was tasked with (undertook) the goal of actively maintaining it.[45] One of the key points of *Urban Sprawl, Global Warming, and the Empire of Capital* is that the U.S. government relied on its domestic stocks of fossil fuels in an effort to resuscitate the American economy from the Great Depression. Subsequent research shows that this approach of using urban sprawl as an economic stimulus was initialized in the early 1920s after an economic downturn following World War I (Chapter 4 of this book). Consumers in the United States (not counting businesses and government) today are the leading purchasers of the world. They take in roughly 20 percent of the globe's total production of goods and services.[46] Therefore, urban sprawl and the demand it creates for consumer durables (retail items expected to last at least three years [e.g., automobiles]) in the United States have profound implications for the capitalist world system and its stability/viability.[47]

Thus, energy supply and national security in the United States are not linked in the conventional sense of minimizing energy use in an effort to shield the domestic economy from the exporting decisions of nations that control surplus energy.[48] (This is the approach that the countries of Western Europe and Japan adopt.[49]) Instead, the proximate concern of the U.S. government as it relates to energy is propping up what Ellen Meksins Woods labels the "empire of capital"[50] through urban sprawl. Put differently, urban sprawl in the United States has been a center of gravity for the American-led world system (i.e., for the American Empire[51])—drawing in allies with access to the economic demand created by urban sprawl[52] and punishing/destroying adversaries by denying access.[53] Thus, oil depletion is not simply an economic phenomenon, but implies the end of U.S. global empire and the world political system as we know it.[54] In other words, the American global system has been predicated on surplus petroleum, and it cannot persist in the absence of this surplus or without some other surplus energy that can economically power urban sprawl. It is this need for surplus power that is ostensibly driving the nuclear energy revival in the United States. (It also explains why there has been an increase in spending on solar power [and biofuel[55]] by the U.S. government.[56]) Thus, energy supply politics in the United States is the politics of global hegemony.

The promotion of civilian nuclear power in the 1950s by the United States was also about global hegemony, but in different ways than it is today. If the United States' allies became dependent on nuclear energy, then American dominance of this technology would reinforce its economic/political dominance.[57] The U.S. domestic deployment of nuclear power plants in the 1950s, 1960s, and 1970s would allow the perfection of such technology and a demonstration of its effectiveness.

In the immediate post–War World II period, the United States sought to limit its Western European allies' knowledge of civilian nuclear power by subsidizing/directing Western European scientists' research in this area. Through such subsidies, Western European scientists were drawn into nuclear energy (and not into solar). The United States, however, tried to circumscribe these scientists' activities to abstract research on nuclear power and not its application. The United States attempted to maintain a monopoly on the application of civilian nuclear energy by classifying all of its knowhow on this matter.[58]

In analyzing the seeming demise of civilian nuclear power in the United States during the late 1970s, researchers emphasize the role of domestic political opposition, safety issues, and economics.[59] American power companies stopped ordering nuclear power plants domestically, however, when it became clear that the United States no longer dominated/controlled the enrichment of nuclear fuel.[60] (The enrichment of nuclear fuel refers to the process of increasing in this fuel the amount of uranium 235 [^{235}U]—the most readily/easily fissionable kind of uranium.) Political scientist Joseph Camilleri, in *The State and Nuclear Power*, points out that "for a great many years the United States enjoyed a monopoly of commercial enrichment capacity within the western world," and that "during the 1970s more than 90 percent of the world's nuclear power stations were fueled with enriched uranium."[61]

In an effort to entice private firms into the enrichment of nuclear fuel, the Nixon administration (1969–1974) drove up the price of enriched fuel.[62] This prompted governments in other countries to aggressively develop their own enrichment capacity.[63] Shortly thereafter, U.S. nuclear power companies realized that to remain competitive in selling power plants internationally, they had to offer countries nuclear fuel processing technology.[64] The Carter administration (1977–1981) prohibited firms from exporting such technology[65] because the processing of nuclear material can be used to substantially expand the amount of weapons-grade material.[66] Once other countries actively deployed enrichment technology,[67] civilian nuclear power was no longer a lever the United States could potentially use to economically/politically dominate others.[68] Instead, it became a major concern with regard to nuclear weapons proliferation.[69]

Why Not Solar?

Why did the United States forego solar in the 1950s? Solar does not ostensibly have the potential of being an imperial tool, like nuclear did. Most obviously, no one country can dominate the source energy of solar (i.e., the sun, wind, waves)—as opposed to nuclear, in which raw uranium and processed fuel are subject to strategic control.

Additionally, nuclear power is based on a highly complex and centralized technology (i.e., nuclear energy plants).[70] This creates potentially high barriers to entering the nuclear energy field. Therefore, if the United States had fully solved the nuclear civilian formula, this knowledge could have been used to dominate the global energy system. (One factor prompting the U.S. government into civilian nuclear power was the concern that the Soviets would solve this formula and be able to draw away the United States' allies as a result.[71]) By contrast, solar energy (in all its forms) is a more diffuse/decentralized power source—with energy being captured where it can be found (e.g., deserts, windy canyons, oceans). In the most idealized form, solar is a local electricity source—with power collected on every rooftop, from every body of water with a current, off of every building corner with wind regularly gusting, and so forth.[72]

The success of solar is in reducing the per-unit costs for energy-collecting equipment (i.e., solar panels, photovoltaic cells, wind and wave turbines).[73] This is in contrast to nuclear, in which the lower per-unit energy costs can ostensibly come only from building larger and larger power plants. Perhaps more importantly, nuclear safety can seemingly be ensured only through such centralized plants.[74] This means that capturing solar is potentially simple (i.e., the purchase of energy-collecting equipment), whereas the high capital costs of producing nuclear energy create virtually insurmountable barriers to generating electricity from nuclear reactions.

As a result, a nuclear energy system is easier to dominate by a few actors/investors (i.e., monopolize). Thus, today we see three dominant firms in the field of building nuclear power plants—Russia's Rosatom, France's Areva, and Westinghouse (now owned by Toshiba of Japan).[75] Even when the United States was actively building nuclear power plants in the 1960s and 1970s, there were two dominant American builders of nuclear reactors—General Electric and Westinghouse.[76]

In the case of solar power, currently there is concern that Chinese firms will come to dominate the manufacture of collecting equipment (e.g., solar panels and wind turbines). (China is believed to build over half of all solar collection equipment.) This potential dominance results from China's low labor costs, subsidized loans for solar equipment manufacturers, and free land and copious research funds from the government. The Chinese government also subsidizes the building/operation of domestic solar power collectors.[77]

Nevertheless, there are numerous firms competing in the solar energy manufacturing field, and entrance is possible.[78] The U.S. government does offer subsidies for solar projects—including for the manufacturing of solar collecting equipment.[79] Most significantly, the federal government recently issued $16 billion in loan guarantees for twenty-eight renewable-energy projects.[80] However, in 2010 the *New York Times* reported that several U.S. solar energy projects were in abeyance because of a lack of financial support.[81]

Conclusion

At the dawn of the twenty-first century, the world faces the twin dilemmas of global warming and fossil fuel depletion (particularly petroleum). These dilemmas were in significant part created by three decisions undertaken by the United States in the post–World War II period.[82] The first was to accelerate the sprawling of its urban zones to stabilize its own economy and those of its Cold War allies. This meant that the capitalist camp was predicated on large amounts of surplus energy, which are necessary to maintain and foster energy-profligate urban sprawl.

A second key energy-related decision undertaken in the United States during the 1950s was its pursuit of nuclear power as the prime alternative to fossil fuels. By the late 1970s, it was brought into sharp relief that civilian nuclear power has several debilitating liabilities: It is dangerous to operate; it creates highly hazardous and long-lasting radioactive waste; it is uneconomic; it invokes strong political opposition; and civilian nuclear energy opens the door to greater and greater proliferation of nuclear weapons. Therefore, more and more reliance on nuclear power poses a grave threat to human civilization. As a result, the United States stopped ordering new nuclear power plants, and this energy source seemed destined to fade from existence.[83]

The U.S. decision in the 1950s to back nuclear energy was accompanied by the decision not to make a major public investment in solar energy. The result is that solar energy science and engineering are substantially behind where they otherwise would be. With the threats of global warming and fossil fuel exhaustion, nuclear power made a political comeback—with Germany, for instance, extending the legal life of its seventeen nuclear power plants by seventeen years in 2010[84] and the United States planning the construction of fourteen new reactors.[85] Nevertheless, given the liabilities of nuclear power (e.g., the Fukushima Daiichi nuclear disaster), the world remains dangerously and unsustainably dependent on fossil fuels. The central argument of this book is that economic elites are directly behind all three decisions outlined above. In *Urban Sprawl, Global Warming, and Empire of Capital*, I highlight how economic elites politically sponsored urban sprawl in the 1930s. Subsequent research has shown that this sponsorship was evident in the 1920s (i.e., the President's Conference on Unemployment) (Chapter 4). By the 1950s, when the policy of urban sprawl was well entrenched, prompting the United States to consume profligate sums of fossil fuels, economic elites (particularly Lewis Strauss) backed major federal subsidies for civilian nuclear power. Economic elites also rejected public subsidies for solar energy (Chapter 3). In Chapter 5, I describe how economic elites backed profligate petroleum use and how the United States continued its excessive oil consumption even after the petroleum shocks of the 1970s. By contrast, the countries of Europe sought to curb their exposure to the world fossil fuels market, in part with the expansion of nuclear power (just when the United States was winding down its nuclear-building efforts).

The United States continues to reject conservation as a strategy to cope with the crises of global warming and energy/oil depletion.[86] Instead, the global economic elite advocates technology (i.e., ecological modernization) in response to these mortal environmental crises. The

U.S. and the global commitment to addressing climate change and fossil fuel depletion through alternative fuels and technology are evident with the Asia-Pacific Partnership for Clean Development and Climate, the US-China Clean Energy Forum, the World Business Council for Sustainable Development (WBCSD), and the International Chamber of Commerce (ICC) (Chapter 6). There are no current technological solutions to the climate change and energy crises (at least not without triggering a nuclear-related crisis [i.e., excessive waste and potential weapons proliferation]), and as a result, there is no global strategy to address climate change or energy supply issues (e.g., the failure of the 2009 Copenhagen Climate Change Conference).

In the next chapter (Chapter 2), I outline different theories that can be used to analyze state behavior and, more specifically, the energy policies that are the subject of this book. The two central theories that I posit are state autonomy theory and economic elite theory. Proponents of state autonomy theory hold that government elites (or political elites) are the main drivers of state behavior. Advocates of economic elite theory contend that economic elites are at the center of public policy formation.

Chapter 2

The Political Economy
of U.S. Energy Policy

In Chapter 1, I outlined how U.S. energy policy is at the center of its grand strategy. On the demand side, the United States consumes massive sums of energy (including petroleum) to maintain a domestic system of urban sprawl in order to sustain the capitalist world system it leads. Going back to the 1950s, the United States pursued civilian nuclear power in an effort to dominate the global energy infrastructure. (U.S. policies in the Middle East [e.g., the Iraq invasion (Chapter 6 of this book)] and Central Asia [e.g., the Afghan War] serve this latter goal.)

A central question of this book is, who determines U.S. energy policy? Two theories predominate in terms of identifying the actors who underlie U.S. state behavior. One is state autonomy theory, and the other economic elite theory. Proponents of state autonomy theory argue that officials within government predominate in the policy formation process. Moreover, these officials will draw in public interest advocates (e.g., environmental groups, economists, energy experts) and their ideas into policy making. Those who hold economic elite theory contend that economic elites operate through policy networks that are financed and led by them. These elites are then able to incorporate ideas and proposals generated within these networks into state policy.

Sometimes economic elite political dominance will manifest itself as special interest politics, and in different instances economic elites will coalesce around ideas and policy proposals that affect the corporate community as whole (i.e., reflect the general interest of the economic elite as a whole). Corporate special interest politics is recognized as plural elitism.

Energy Policy as Special Interest Politics

Urban Sprawl

A salient argument deployed to account for U.S. urban sprawl is that particular economic interests (e.g., large landholders) have been successful in instituting pro–urban sprawl polices in order to benefit their bottom line. This view of the politics of urban sprawl is consistent with the plural elitism take on policy making in the United States. Plural elitism grew out of the pluralism theory of the policy-making process.

Pluralism arose as the dominant political science paradigm in the post–World War II period. Pluralist theorists, most prominent among them being Robert Dahl, hold that various interest groups, including major corporations and labor unions, exercise influence over government.[1]

The near total dominance of the theory of pluralism in American political science ended in the late 1960s and early 1970s with the social movements of this period (e.g., the antiwar movement, the civil rights movement, and the environmental movement). What came into full relief during the later 1960s and early 1970s was that government was not a neutral arena whereby different interest groups brought their political resources (i.e., money, votes, prestige) to bear, as maintained by early pluralist thinkers. Nor was the successful mobilization of interest groups all that was needed to influence/shape the policy-making process.[2] Instead, political influence in the United States came to be viewed as consistent with plural elitism.

Plural elitism theorists hold that certain interests are entrenched and exercise dominant influence over policy formation. Theodore Lowi explains that the allocation of policy-making authority to specific agencies within the executive branch leads to the "capture" of those agencies by special interests—and thus the establishment of what he calls "subgovernments." The practice of ceding policy-making authority to executive branch agencies is named by Lowi "interest-group liberalism."[3]

Grant McConnell, like Lowi, attributed the diffusion of state power to a dominant political philosophy. This political philosophy, according to McConnell, is rooted in discourses developed during the Progressive Era. These discourses posit that democracy is most effectively applied in small bureaucratic units. In turn, this fracturing of the federal government into a multitude of small units allows the capture of significant amounts of state power by special interests.[4] Hence, while both McConnell and Lowi trace the public philosophy that has predicated the creation of a governmental structure that promotes capture by special interests to different philosophical precepts, their conclusions are similar.

Whereas Lowi and McConnell attribute the creation of subgovernments to the institutional structure of the federal government, especially the executive branch, and the legislative practice of delegating policy-making authority to executive branch agencies, Dahl and Lindblom, in a 1976 modification of early pluralist thought, argue that business groups in particular are going to have privileged access to the policy-making process. Subgovernments, they aver, are less the result of happenstance and more the result of the fact that businesspeople are directly responsible for running the economy. The result of this responsibility is the "privileged participation of business" in government:

> Businessmen are not ordered by law to perform the many organizational and leadership tasks that are delegated to them. All these societies operate by rules that require that businessmen be induced rather than commanded. It is therefore clear that these societies must provide sufficient benefits or indulgences to businessmen to constitute an inducement for them to perform their assigned tasks.
>
> The consequence of these arrangements—peculiar as they would appear to a man from Mars—is that it becomes a major task of government to design and maintain an inducement system for businessmen, to be solicitous of business interests, and to grant to them, for its value as an incentive, an intimacy of participation in government itself. In all these respects the relation between government and business is unlike the relation between government and any other group in the society.[5]

Therefore, subgovernments are the logical outcome of an economic system that relies on private elites to deliver economic prosperity. Giving businesspeople dominant influence over those government agencies that shape the behavior of the economy helps to ensure that the policies of these agencies will lead to economic growth and stability. Arthur Selwyn Miller refers to this arrangement as the "fusion of economic and political power."[6]

Dahl's and Lindblom's argument that political authority over economic policies must be ceded to economic interests in order for those policies to be successful is consistent with the history of the Federal Housing Authority (FHA).[7] The FHA was given responsibility over the federal government's prime housing program beginning in the 1930s, and policy-making positions within the FHA were granted to prominent individuals from the housing industry as well as from the financial sector.[8] As indicated by Dahl and Lindblom, the fact that individuals with such backgrounds were given responsibility to set the federal government's housing policy is logical because it was the housing industry and the financial sector that were ultimately entrusted with building and financing the nation's housing, even that housing that was sponsored by the FHA. From the post–World War II period into the late 1960s, the FHA played the key governmental role in subsidizing and encouraging urban sprawl in the United States.[9]

U.S. international oil policy in the 1920s is also consistent with plural elitism. Joan Hoff Wilson, in her history of U.S. foreign policy during this period, found that American petroleum companies were ceded the authority to negotiate oil agreements with other countries on behalf of the United States.[10]

The profound global implications of urban sprawl in the United States challenges the notion that urban sprawl in the United States is solely a function of special interest politics. Today, urban sprawl remains as a means to prop up the world economy. To this end, petroleum and natural gas supplies in the Middle East and Central Asia remain as key prizes.

Civilian Nuclear Power

There is a wealth of literature that casts U.S. civilian nuclear power policies as a product of the nuclear industry itself.[11] Perhaps the best and most widely-cited example of this literature is Mark Hertsgaard's

Nuclear Inc.: The Men and Money behind Nuclear Energy.[12] Hertsgaard argues that the leadership of the U.S. nuclear power industry is integrated into what he refers to as the Atom Brotherhood, and because of the finite nature of fossil fuels and the "greenhouse effect,"[13] this brotherhood foresees the inevitability of a nuclear-powered America (and world). Through its deep pockets as well as corporate and political connections, nuclear reactor manufacturers have garnered huge sums in subsidies (for research, including demonstration nuclear power plants) and have been extended preferential protection from lawsuits resulting from the accidental release of radioactivity.[14] Moreover, the government directly took on the task of enriching (and recycling[15]) nuclear fuel, as well as taking ultimate responsibility for the long-term storage of nuclear waste.[16] Hertsgaard wrote his seminal book in 1983, when the U.S. nuclear power industry still appeared capable of recovering in the short term from the setbacks of the late 1970s (most prominently the Three Mile Island accident).

Rick Eckstein's *Nuclear Power and Social Power* identifies a different component of the special interest politics surrounding nuclear power in the United States.[17] Eckstein specifically points to local growth coalitions as potential opponents to nuclear power plants.[18] According to Eckstein, it was local business interests that successfully defeated the start-up of the completed Shoreham nuclear power plant on Long Island, New York. Nuclear plants are potential threats to public health and, hence, to local real estate values, as well as to the local business climate, as people and firms can be reticent to locate in the vicinity of such a plant. In spite of the potential opposition that Eckstein notes, in the contemporary period the federal government has reinitiated nuclear power plant building.

U.S. policies on nuclear power cannot be the sum of special interest politics. Most glaringly, there is no evidence that the politically potent fossil fuel industry (i.e., oil, natural gas, and coal) has ever politically stood in the way of the federal government's development and promotion of nuclear power, even though a hugely successful civilian power program would have eliminated the need for fossil fuel as a source of energy (especially coal). Although nuclear power is not a direct economic threat to petroleum (i.e., gasoline) for the powering of automobiles, nuclear power in theory held the potential of generating so much cheap surplus energy that electrically powered automobiles, or those propelled by hydrogen, could be feasible.[19]

Far from manifesting opposition, the fossil fuel industry demonstrated significant political support for nuclear power in the 1950s. This support was shown through the 1956 report *Peaceful Uses of Atomic Energy*, aka the McKinney Report, which was submitted to the Congressional Joint Committee on Atomic Energy. The report was compiled by the Panel on the Impact of the Peaceful Uses of Atomic Energy. To write its report, the panel drew upon "qualified individuals, organizations and study groups, each operating autonomously and submitting their independent findings of fact and their conclusions to seminar discussion groups. . . . All in all, 327 people, all authorities in their field, took part in this work."[20] As outlined in Chapter 3, this panel recommended public financial support for civil nuclear power. Numerous fossil fuel firms and trade associations helped write the panel report. Among them were the American Petroleum Institute (trade association), American Gas Association (trade association), Appalachian Coals, Inc., Gulf Oil, National Coal Association (trade association), National Petroleum Council (trade association), Shell Oil; Texas Company (oil firm), Standard Oil of California, Standard Oil of Indiana, and Standard Oil of New Jersey.[21]

The panel showed indifference to the adverse impact that a nuclear power program could have on the fossil fuel industry. It wrote: "Disruptive influences, even on specific industries most directly affected [by nuclear power], are likely to come—if at all—over periods of time long enough to permit orderly adjustment." The report goes on to state that with regard to those "specific industries popularly assumed to be most vulnerable to atomic inroads—coal, for example—such dislocations as appears possible would come from a welter of forces more complex and more overriding than atomic energy alone." The panel adds that "if atomic power is exploited as a source of electric power at a rate consistent with sound technological, economic and public policy considerations, the impact will be totally beneficial at home and abroad."[22]

Therefore, neither the federal government's policies on urban sprawl nor nuclear civilian power can be accounted for by simply looking at special interest politics. Instead, we must look at those political processes in which the general interests of the U.S. polity and economy are identified and acted upon. As noted above, there are two different theoretical camps that seek to identify and analyze the processes whereby the general political interests of the United States are formulated and implemented: 1) state autonomy theory and 2) economic elite theory.

State Autonomy Theory

At the core of state autonomy theory is the notion that officials within the state can and do behave autonomously of all social groups.[23] Officials within the state have special theoretical significance because they are often looked upon to deal with political and economic matters. Moreover, they are provided in many instances with the resources, such as legal authority and a budget, to do so.[24] Indicative of the argument that autonomous officials within the government drive state behavior, Adam Rome, in his book linking the rise of modern environmentalism in the United States to urban sprawl, holds that the federal government beginning in the 1920s viewed low-density housing development as the means to attain broad-based home ownership.[25] Also consistent with the state autonomy position is Stephen Krasner's argument that U.S. foreign policy as it relates to raw materials, including petroleum, has historically been shaped by the ideology of officials within the state.[26]

In this context, autonomous policy makers can and do draw upon different public interest advocates, scientists, and economists to determine how to prioritize various issues and how to address them.[27] In this way, public interest groups, for instance, are incorporated into the policy-making process. Scientists and economists have specific importance within state autonomy theory. This is because they offer the technical know-how to instruct public officials. Scientists and economists also orient state officials to the political, economic, environmental, and social issues that must be addressed in order to avoid more serious difficulties.[28] According to Theda Skocpol, the legitimacy and usefulness of experts (i.e., scientists and economists) is enhanced by the fact that they "most often . . . attempt to act as 'third-force' mediators, downplaying the role of class interests and class struggles and promoting the expansion of state or other 'public' capacities to regulate the economy and social relations."[29]

On the question of climate change, however, the federal government historically has shunned the advice and activism of numerous scientists and environmental groups.[30] A strong consensus has developed among scientists that the continuing uncontrolled emission of carbon dioxide holds seemingly dire consequences for the earth's biosphere. This consensus includes the Intergovernmental Panel on Climate Change, a panel composed of leading climate scientists.[31] The Clinton administration did sign the 1997 Kyoto Protocol, which is geared toward abating greenhouse gases. Nevertheless, it did so reluctantly, and the administration demanded a number of loopholes for the United States in the implementation of the protocol. As a result of U.S. efforts to gain exemptions for itself in the application of the protocol, an agreement to implement the protocol could not be negotiated. In early 2001, the second Bush administration withdrew from the protocol.[32] With the United States withdrawn from the protocol, a sufficient number of the remaining signatories successfully negotiated its implementation, and among these countries the protocol went into effect in 2005.[33]

In the 2009 Copenhagen Climate Change Conference, the Obama administration was unable to reach agreement with other governments of the world on the issue of a binding global warming agreement. At the heart of this failure was the United States' inability/unwillingness to specify how it would address its massive greenhouse gas emissions (with the U.S. Senate not passing climate change legislation on the eve of the conference).[34]

On the issue of civilian nuclear power, the scientists who advised the U.S. Atomic Energy Commission (AEC) in the late 1940s recommended against nuclear-generated electricity. The commission's scientists held that the science of civilian nuclear power was too uncertain to employ in the immediate term (Chapter 3 of this book).

Other scientists argued on behalf of solar energy—most famously through the 1952 presidential Paley Commission report.[35] While acknowledging the expense and relatively low energy intensity of solar power, advocates of solar power in the 1950s and 1960s nevertheless asserted that government should fully research the potential of solar for electricity generation. The finite nature of fossil fuels seemingly necessitates the aggressive funding of such a research agenda.[36] In spite of such advocacy and logic, political scientist Frank Laird, in assessing energy policy from the Truman through the Johnson administrations, concludes that "solar never got serious consideration as a source that might be a major part of the future U.S. energy system and so it lacked the support that such a role might have entailed."[37]

Historians of U.S. solar research policy Harvey Strum and Fred Strum note that "even after the 1973 oil crisis solar energy did not obtain funding in proportion to that given to nuclear and fossil fuels. In 1974 the federal government spent $15 million on solar research and in 1975 spent $30 million."[38] Federal solar policy was hobbled by a pro-nuclear bias within federal energy research agencies and by the opposition to certain types of solar research by the electric utility industry.[39] The utility industry was opposed to the development of solar power technology that would allow consumers to generate their own electricity.[40] It was only during the last two years of the Carter administration (1979–1980) that solar research attained political and financial priority.[41] These years coincide with the oil crisis brought on by the Iranian Revolution. Given the special interest politics histori- cally surrounding solar power, Strum and Strum, writing in 1983, conclude that "with the exception of the last two years of the Carter Administration, American energy policy has ignored the potential of solar energy while relying on nuclear energy as the only real alterna- tive to fossil fuels."[42]

Economic Elite Theory

Whereas plural elite theorists describe how individual corporate deci- sion makers dominate specific and narrow policy areas,[43] economic elite theorists contend that these corporate decision makers, along with other individuals of wealth, develop and impose broadly construed policies on the state. Additionally, whereas plural elite theory views the business community as socially and politically fragmented, proponents of the economic elite approach hold that the owners and leadership of this community can be most aptly characterized as composing a coherent social and political unit or class.[44]

Clyde Barrow points out that "typically, members of the capital- ist class [or the economic elite] are identified as those persons who manage [major] corporations and/or own those corporations." He adds that this group composes no more than 0.5 to 1.0 percent of the total U.S. population.[45] This group as a whole is the upper class and the upper echelon of the corporate or business community. The resource that members of the economic elite possess that allows them to exercise a high level of influence over government institutions is wealth. The wealth and income of the economic elite allow it to accumulate superior amounts of other valuable resources, such as social status, deference, prestige, organization, campaign finance, lobbying, political access, and legal and scientific expertise.[46]

According to the economic elite approach, despite the segmentation of the economic elite along lines that are related to their material holdings, most policy differences that arise because of differences in economic interests can be and are mediated. Social and organizational mechanisms exist that allow business leaders to resolve difficulties that develop within a particular segment and between different segments of the corporate community. For specific industries, or for disagreements between different industries, trade or business associations can serve as organizations to mediate conflict. William Appleman Williams, in his extensive history of the U.S. politics during the nineteenth century surrounding economic, foreign, and trade policies, explains that agricultural interests throughout the country formed business associations to address their common problem: how to gain access to new markets to profitably absorb the agricultural surpluses produced in the United States. Williams writes:

> [Agribusinesses] participated in the general movement to create agricultural clubs and societies. Whether formed on a national scale, like the American Shorthorn Association (1846) and the Agricultural Society (1852), or organized on a state basis, like the Indiana Horticultural Society (1841–1842) and the Wisconsin Agricultural Society (1851), such groups . . . totaled 621 by 1849.[47]

Williams notes that these organizations "helped ease some" of the regional, economic, and political "conflicts" that emanated from the immense and varied U.S. agricultural sector.[48]

Social institutions, such as social and country clubs, can also serve as a means through which to develop political consensus among the upper echelon of the business community on various economic, political, and social issues.[49] Michael Useem, based on his extensive study of large American and British corporations, argues that corporate directors who hold membership on more than one board of directors tend to serve as a means through which the corporate community achieves consensus on various political issues.[50]

On broad issues such as urban sprawl and civilian nuclear power, business leaders are also able to arrive at policy agreement and consensus through "policy-planning networks." According to G. William Domhoff, the policy-planning network is composed of four major components: policy discussion groups, foundations, think tanks, and university research institutes. This network's budget, in large part, is drawn directly from the corporate community. Furthermore, many of the directors and trustees of the organizations that comprise this policy-planning network are often drawn directly from the upper echelons of the corporate community and from the upper class. These trustees and directors, in turn, help set the general direction of the policy-planning organizations, as well as directly choose the individuals who manage the day-to-day operation of these organizations.[51]

Domhoff describes the political behavior of those members of the economic elite who manage and operate within the policy-planning network:

> The policy-formation process is the means by which the power elite formulates policy on larger issues. It is within the organizations of the policy-planning network that the various special interests join together to forge, however, slowly and gropingly, the general policies that will benefit them as a whole. It is within the policy process that the various sectors of the business community transcend their interest-group consciousness and develop an overall class consciousness.[52]

Therefore, those members of the economic elite who operate within the policy-planning network take on a broad perspective and act on behalf of the economic elite as a whole. Within this policy-planning network, members of the economic elite are interested in general positions on such issues as foreign policy, economic policy, business regulation, environmental policy, and defense policy questions.[53] David A. Wells, a well-known figure in business and political circles during the last third of the nineteenth century, argued that the growing U.S. industrial base needed access to external markets in order to maintain stability and profitability. In a similar vein, Captain Alfred T. Mahan, during the same period, famously held that the United States needed to greatly enhance its naval capacity in order to secure vital shipping lanes to foreign commercial markets.[54] As described in Chapter 3, the Rockefeller Foundation in the 1930s provided the initial financial support for the research of nuclear energy. In Chapter 6, I explain how a neoconservative policy group, the Project for a New American Century, led by economic elites Donald Rumsfeld and Richard "Dick" Cheney, argued in the late 1990s for the invasion of Iraq.

This broad perspective also allows the policy-planning network to develop plans and positions to deal with other groups and classes. The network, for example, develops positions and plans concerning such policy areas as welfare and education. These plans can take several forms depending on the scope and level of the problems facing the business community and the state.[55]

Domhoff argues that the focal point in the policy-planning network is the policy discussion group. The other components of the policy-planning network—foundations, think tanks, and university research institutes—generally provide original research, policy specialists, and ideas to the policy discussion groups.[56] Policy discussion groups are largely composed of members from the corporate community and the upper class. Examples of policy discussion groups are the Council on Foreign Relations, the Committee for Economic Development, the National Association of Manufacturers, and the U.S. Chamber of Commerce. Overall, policy discussion groups are the arenas where members of the economic elite come together with policy specialists to formulate policy positions and where members of the economic elite evaluate policy specialists for possible service in government.[57] Two other prominent examples of corporate policy discussion groups are the WBCSD and the ICC. As I explain in Chapter 6 of this book, both of these organizations are composed of firms from all over the world. In the face of contemporary concerns about waning energy reserves, especially petroleum, these groups do not advocate energy conservation by reducing urban sprawl. Instead, the WBCSD and the ICC stress energy efficiency and the development of alternative sources of energy (including nuclear).

Certain environmental groups, in terms of their leadership and/or financing, have the characteristics of economic elite–led policy-planning organizations. These groups include the Sierra Club prior to the 1960s, Save-the-Redwoods League, and the Environmental Defense Fund. The Environmental Defense Fund, for instance, receives significant financing from large foundations, and it has several corporate executives on its board of directors.[58] Susan R. Schrepfer, in her survey of the

Sierra Club's early charter members, found that approximately one-third were academics, and "the rest of them were almost all businessmen and lawyers working in San Francisco's financial district."[59] The club was founded in 1892. Schrepfer goes on to explain that businesspeople continued to compose a substantial portion of the club's membership and leadership until the 1960s.[60] Unlike the Sierra Club, the high level of economic elite participation on Save-the-Redwoods League's governing council has been maintained throughout its history. The closed governance structure of the league created the "tendency for the council and board to be increasingly dominated by businessmen and patricians, while fewer academics were drawn into the organization's leadership in the 1950s and 1960s."[61]

Economic elites can use their relationships with environmental and natural resources groups to gain information and policy proposals in their efforts to shape public policies on environmental and natural resource questions when deemed necessary.[62] In 2007, the Natural Resources Defense Council, along with the Environmental Defense Fund, the World Resources Institute, and the Pew Center on Global Climate Change, formed the United States Climate Action Partnership with ten major businesses: Dupont, General Electric, Alcoa, Caterpillar, Duke Energy, PG&E of California, the FPL Group of Florida, PNM Resources of New Mexico, British Petroleum, and Lehman Brothers. The political goal of the partnership is to reduce climate change emissions through the development and deployment of energy-efficient and abatement technologies.[63]

Economic elite–led policy discussion groups have also been formed for the purpose of shaping decision making on the urban level. One prominent example of such an entity is the National Municipal League.[64] From the nationwide effort of this organization came the Progressive Era urban reforms of the civil service "to regulate personnel practices, competitive bidding to control procurement, the city manager form of government to systematize decision making, and at-large elections to dilute the voting power of the working classes."[65]

Returning to the issue of the "general interests" of the capitalist polity and economy, the economic elite approach would suggest that the conceptions of the general interest that dominate the state are not determined within the state in response to different shifts in the operation of the political economy and/or public opinion. This view is implicit in the neo-Marxist view of politics[66] as well as in state autonomy theory. Instead, it is economic elites and producer groups, operating through policy-planning networks, that determine which issues within capitalism are to be addressed by the state and how.

Locally oriented economic elites (e.g., large landowners, land developers, owners of utilities and local media outlets, and real estate attorneys) have historically imposed the objective of local economic growth on local and state governments in order to inflate land values and expand the local consumer base. Together, these particular elites have been labeled local growth coalitions by Harvey Molotch.[67] In the United States, it was locally oriented economic elites (especially large landowners and developers) who developed the techniques and impetus for early urban sprawl. The techniques of urban sprawl were spread and standardized through economic elite–led policy-planning groups—most prominently, the Home Builders and Subdividers Division and the City Planning Committee of the National Association of Real Estate Boards (NAREB).[68] In Chapter 4, I outline how, as the broad economic benefits of urban sprawl became apparent to economic elites in the 1920s, the federal government began promoting urban sprawl.

Conclusion

Among countries with a population of more than 35 million, the United States is the largest per capita consumer of energy as well as by far the largest per capita emitter of carbon dioxide (the key greenhouse gas).[69] Thus, it plays a particularly significant role in the questions of global warming, energy depletion, and the failure of the international politics of climate change.

What is truly amazing (as well as frustrating) is that the U.S. government is unwilling to actively employ conservation measures (e.g., hefty gasoline taxes[70]) as a response to the twin potentially catastrophic crises of energy depletion and global warming.[71] It is this inability/ unwillingness of the United States to curb its energy consumption that is ostensibly a prime obstacle to an effective worldwide global warming abatement regime. (It is unfeasible to expect any country to significantly sacrifice wealth creation and/or national sovereignty to address global warming unless the United States cuts back its huge per capita carbon dioxide emissions.) In *Urban Sprawl, Global Warming, and the Empire of Capital*, I argue that the United States is unable to employ policies to aggressively reduce its energy consumption (and greenhouse gas emissions) because urban sprawl in the United States plays a substantial role in propping up the global economy. In Chapter 4 of this book, I outline how this positive economic role for urban sprawl was identified in the United States during the 1920s.

In lieu of the United States directly addressing its massive energy consumption and greenhouse gas emissions, alternative energy (including nuclear) has become a glaring issue on the U.S. (and world) agenda. This brings the history of U.S. energy investments into prominence. Among energy sources other than fossil fuels, the U.S. government has historically prioritized nuclear power. The United States' unwillingness to aggressively research solar energy means that it and the world do not know the full potential of this now-vital energy source.

Scholars have sought to explain U.S. energy demand and supply policies in terms of special interest politics. Urban sprawl and nuclear civilian energy in the United States, however, were not simply the product of special interest projects. Instead, both were the result of a particular conception of the United States' national and global interests. My argument is that this conception was formulated by economic elites through policy-planning networks and imposed on the state by them.

Chapter 3

U.S. Economic Elites, Nuclear Power, and Solar Energy

Historically, U.S. urban sprawl and U.S. nuclear policy have been linked, but not in the way one might think. As explained in the next chapter, urban sprawl in the 1920s was embraced as a means to stimulate the economy. Nuclear power in the 1950s was not necessarily intended to meet the growing energy demand created with the sprawling of the United States' urban zones. Instead, nuclear energy was initially developed by the United States as a hegemonic policy. The link between urban sprawl and nuclear power is that they were initiated and politically sponsored in the United States by economic elites.

U.S. economic elites championed nuclear power. In addition to the efforts of AEC chairperson and investment banker Lewis Strauss, economic elites supported nuclear energy through the policy discussion groups of the Rockefeller Foundation and the Panel on the Impact of the Peaceful Uses of Atomic Energy. Conversely, at about the same time that the Panel on the Impact of the Peaceful Uses of Atomic Energy was strongly advocating for public financing of nuclear power, the economic elite leaders of the Association for Applied Solar Energy (AFASE) (based in Arizona) in 1955 expressed opposition to government support for solar energy.

U.S. Economic Elites and Nuclear Power

In a 1956 report, the Rockefeller Foundation leadership outlines the foundation's early and leading role in the development of nuclear science:

29

The Rockefeller Foundation had . . . taken a lively interest
in nuclear research . . . [I]ts funds had provided fellowship
assistance to many whose prepared minds were to play a
significant role. Among those . . . were such scientists as
Robert F. Bacher (1930–32), Hans Bethe (1930–32), Arthur
Compton (1919–20), Edward O. Lawrence (1925–27), J. R.
Oppenheimer (1927–28), Henry De W. Smyth (1921–24),
Edward Teller (1933–34), and John A. Wheeler (1933–35).

The authors of 1956 report go on to note:

In addition to opportunities for further study by individuals,
Foundation funds assisted a number of laboratories with
buildings, such items of equipment as electrostatic genera-
tors, cyclotrons, and betatrons, and free research funds for
nuclear investigations. One notable group of laboratories was
at the University of Copenhagen, where the physicist Niels
Bohr, the chemist George von Hevesy, and the physiologist
August Krogh led a distinguished company of scientists in
pooling the resources of their several disciplines to work at
such questions as the biological uses of isotopes. Another
was the Radiation Laboratory of the University of California at Berkeley, where Ernest O. Lawrence devised and
rapidly developed the cyclotron. The list would include the
Collège de France, the University of Minnesota, Rochester,
Stockholm, Washington University at St. Louis, the Mas-
sachusetts Institute of Technology, Columbia, Chicago,
Princeton, the University of Sao Paulo. Two of the last
grants made by the Foundation before the field of nuclear
research was swept up into the wartime Manhattan Project
provided $60,000 in 1942 to expedite the winding of the
armature of the giant magnet of Lawrence's new 184-inch
cyclotron, and a sum of $100,000 at about the same time to
the Metallurgical Laboratory of the University of Chicago
for research in problems of industrial hygiene arising from
the handling of radioactive materials.[1]

The Rockefeller Foundation was founded in 1909 with a $50
million endowment by John D. Rockefeller, Sr., and has been led by

the Rockefeller family ever since.[2] The foundation has been a leading financial sponsor of scientific and public policy research throughout the twentieth century and into the contemporary era.[3] Edward H. Berman, in his history of the relationship of the Carnegie, Ford, and Rockefeller Foundations to America's international relations, writes that these "foundations have consistently supported the major aims of United States foreign policy, while simultaneously helping to construct an intellectual framework supportive of that policy's major tenets."[4]

Within the Roosevelt administration, atomic research (i.e., the Manhattan Project) was overseen by Vannevar Bush.[5] Bush was a leading scientific administrator for the corporate community. Bush's biographer notes:

> In January 1938, Bush wooed a group of leading corporate researchers at a meeting in New York City, offering his revamped [MIT] engineering division as a forum for the benefit of corporate laboratories. In attendance were nearly 50 executives from some of the nation's leading companies: Procter & Gamble, Champion Paper, Colgate-Palmolive, Swift, Lilly Research, Burroughs Wellcome and Dodge.[6]

In 1939, Bush became president of the Carnegie Institution, which was "founded by wealthy industrialist Andrew Carnegie in 1902. The institution had an endowment of $33 million and spent $1.5 million annually on research." Bush's biographer described the institution's board of trustees as "larded with rich and influential members" including Herbert Hoover, Frederic Delano (Franklin Roosevelt's uncle), and W. Cameron Forbes ("a member of a wealthy Boston family").[7]

The AEC

In the 1950s and into the late 1960s, the institutional support for nuclear power came predominately from the AEC. A federal independent regulatory commission, the AEC funded virtually all research that went into nuclear energy during this period.[8]

Significantly, the AEC's General Advisory Committee advised against civilian nuclear in 1947.[9] This committee (made up of nine members) was "the top scientific" adviser to the AEC.[10] Headed at the time by arguably the foremost nuclear scientist in the world, J.

Robert Oppenheimer, the General Advisory Committee concluded that "it does not appear hopeful to use natural uranium directly as an adequate source of fuel for atomic power." In a revised report, the committee nevertheless held that it did "not see how it would be possible under the most favorable circumstances to have any considerable portion of the present power supply of the world replaced by nuclear fuel before the expiration of twenty years."[11]

Ignoring the advice of its science advisory committee, the AEC, under the chairpersonship of Lewis Strauss, aggressively sponsored nuclear power. Strauss was a wealthy investment banker who by his twenties was "accepted into the prestigious American Banking Association and New York Chamber of Commerce and his name appeared in *Who's Who* beginning in 1924."[12] By the time of his initial appointment to the AEC, Strauss's "wider circle of acquaintances, included nearly all the powerful men who dominated American business, finance, and politics."[13] During his investment banking career, Strauss had become acquainted with the nuclear science community. Strauss's biographer holds that "by the middle of 1939, Strauss knew many of the leading [nuclear] physicists. They accepted him into their fraternity, if not as a fellow scholar, then as more than a mere financier."[14] In 1946, he was among the first five appointees to the AEC.[15] It is noteworthy that when Strauss left the AEC, he went to work for the Rockefeller family as a financial adviser.[16]

In 1953, Strauss was appointed chair of the AEC by the Eisenhower administration. As chair of the AEC, Strauss became a leading champion of nuclear energy. He personally brokered the first nuclear reactor used for the generation of electricity in the United States. As his biographer reports, "Strauss had persuaded Philip Fleger, chief executive of Duquesne Light Company, to provide the site [in Shippingport, Pennsylvania], build the generator to be linked to the reactor, and connect the plant into Duquesne Light's network. . . . The AEC contributed a substantial portion of the capital cost." Strauss's biographer explains that Strauss "saw in [the reactor] an opportunity to demonstrate the potential for nuclear power":[17]

> Already he could see that in Europe, where the cost of producing electricity in conventional power plants was far greater than in the United States, atomic power could become cost-effective more quickly. An American firm had

sold a reactor to Belgium in 1957, and Strauss expected other orders from European countries to follow quickly.

Strauss's biographer adds that "he knew that the design, manufacture, and operation of such plants [e.g., Shippingport] by American firms was the best way to insure the technological advances that would reduce the costs [of nuclear power] to a competitive level."[18] The Shippingport reactor was completed and operating in 1957.

Another prominent economic elite on the AEC in the 1950s was Thomas E. Murray. According to the historians of the commission, Murray was

> a highly successful engineer and business executive in New York . . . and by the time he was appointed to the Commission in March 1950 he had been president of his own company, board member of his family company and several large corporations, trustee of several banks, and a receiver of the Interborough subway system.[19]

In 1953, Murray gave a speech before an "electricity utility convention," where he argued that "attaining economical nuclear power was just as vital to national security as the United States' preeminence in nuclear weapons." Murray went on to assert that "friendly nations were counting on the United States not only to protect them from Soviet aggression but also to supply them with nuclear power technology."[20]

As noted in Chapter 2, during 1956, the report *Peaceful Uses of Atomic Energy* was submitted to the Congressional Joint Committee on Atomic Energy. The report was compiled by the Panel on the Impact of the Peaceful Uses of Atomic Energy. To write its report, the panel drew upon "qualified individuals, organizations and study groups, each operating autonomously and submitting their independent findings of fact and their conclusions to seminar discussion groups. . . . All in all, 327 people, all authorities in their field, took part in this work."[21] Numerous fossil fuel firms and trade associations helped write the panel report. Among them were the American Petroleum Institute (trade association), the American Gas Association (trade association), Appalachian Coals, Inc., Gulf Oil, the National Coal Association (trade association), the National Petroleum Council (trade association), Shell Oil, Texas Company (oil firm), Standard Oil of California, Standard Oil of Indiana, and Standard Oil of New Jersey. Other notable firms/institutions that participated in compiling the panel report included Chase Manhattan Bank, Ford Motor Company, Dupont Chemical, General Dynamics, General Electric, General Motors, Monsanto Chemical, Pacific Gas & Electric, the Rockefeller Foundation, Sullivan & Cromwell (a prominent New York law firm), and the *Washington Post*.[22]

This panel strongly recommended public financial support for civilian nuclear power. It asserted

> that, in the event that industry does not take on the full risks and burdens, the Commission [i.e., the AEC] should support a program to bring atomic power to a point where it can be used effectively and widely on a competitive basis, even to the construction with public funds of one full-scale "demonstration" plant of each major reactor size and type.

The panel pressed that "the urgency associated with this [atomic power] program requires that the technological resources of atomic power be fully explored with *high priority*." The panel's concluding recommendation was that "atomic power be exploited as a source of electric power at a rate consistent with sound technological, economic and public policy considerations." The panel added that "if atomic power is exploited as a source of electric power at a rate consistent with sound technological, economic and public policy considerations, the impact will be totally beneficial at home and abroad."[23]

Under the heading "International Consequences of the Growth of Atomic Power," the Panel speculated that "in the uncommitted areas of the world, American leadership in making atomic power available could be a strong influence in guiding these areas toward the course of freedom" (i.e., the American camp). Thus, "in this sense, atomic power acquires great importance in international relations," and "this consideration should strongly influence our national policy as to the rate at which the development of atomic power suitable for such purposes is pressed." The panel argued that "there is urgency for the development in the United States of atomic power plants suited to the needs of the other nations of the free world."[24] The panel went on to argue that "atomic power may be the most tangible symbol of

America's will to peace through the peaceful atom." Moreover, "If we fail to bring atomic power to the free world, other countries [i.e., the Soviet Union] will do so ahead of us."[25]

General Electric official Everett L. Hollis, writing in 1957 a political/legislative survey of civilian nuclear power in the United States, reported that "the policy that atomic energy be developed as a Government monopoly was to some degree the result of the desire to maintain America's international monopoly." He also reported that "for reasons of foreign policy it was contended that the United States must have a vigorous peacetime atomic program."[26] During the 1950s, 1960s, and 1970s, General Electric was the United States' leading nuclear technology firm.[27]

Solar Power

In contrast to nuclear power, which received substantial research support from the Rockefeller Foundation and later gained the political endorsement of the U.S. corporate community, solar energy was historically starved of institutional help in the United States. This lack of support was devastating to the science, engineering, and knowledge of solar power.

Frank T. Kryza, in *The Power of Light: The Epic Story of Man's Quest to Harness the Sun*, outlines efforts in the late nineteenth and early twentieth centuries to develop industrial-scale solar power projects. Kryza in particular documents Frank Shuman's ingenuity in tapping the power of the sun during the first and second decades of the twentieth century. Shuman was a successful and wealthy Philadelphia inventor who became interested in solar power. He conducted experiments demonstrating the promise of solar power—including the use of water to store ample amounts of heat energy that can be used at night or during cloudy days. Kryza describes the results of Shuman's initial solar power experiments:

> A good-sized coal-fired boiler in the first decade of the twentieth century—the time Shulman was conducting his experiments—would have been capable of generating 100,000 pounds of steam per hour, consuming 2 or 3 tons of coal to do so, to run a 3000-horsepower steam engine.

Shuman's first full-scale demonstration project produced an average of 600 pounds of steam per hour, less than 1 percent of what a large coal-fired boiler might produce. He was tapping from sunlight the amount of energy per hour contained in about 30 pounds of coal. Given this was his first try, Shuman thought these results respectable.[28]

Kryza goes on to note that the problem of running a solar plant at night or during cloudy days "was not difficult to solve":

For 6 hours of every 24-hour cycle, while the sun was shining at its brightest, the heat absorber [or solar collector] deliver water at 212 degrees Fahrenheit to both the steam turbine and the [water] tank. For the remaining 18 hours of the day, the engine would draw off hot water from the surplus stored in the tank, permitting engine operation round the clock.[29]

Kyrza adds that "factoring in liberal figures for heat loss, a tank of the size proposed would still comfortably permit overnight operation of the plant at full throttle—and, with better insulation, even operation during a string of 1 or 2 cloudy days—without interruption."[30]

As Shulman was making gains with his solar experiments, he turned to the "financial oasis that had funded all his earlier ventures—the 'big-money' men of Philadelphia." However, "Pennsylvania was America's premier coal country, and these tycoons were already committed to fossil fuels." Shulman's "former backers reached modestly into their deep pockets, or not at all, to support his new venture. . . . He needed support from people who were not heavily invested in coal."[31]

Shulman was able to find political and financial support from the governments of Great Britain and Germany, both of which were interested in using industrial-scale solar power infrastructure to economically develop their African colonies:

Practical demonstrations of Frank Shulman's solar technology in the years before World War I would win the enthusiastic support of Lord Kitchener of Khartoum, the British proconsul of Egypt; Sir Reginald Wingate, the iron-fisted ruler of neighboring Sudan; and earn Shuman an invitation from the German Reichstag to accept the equivalent in [German marks] of $200,000 in venture capital—a colossal sum equivalent today to millions of dollars—to bring solar power to Germany's growing colonial possessions in Africa.[32]

Shuman's effort at solar-powered irrigation of the Nile River was not entirely successful,[33] but the British and German governments remained committed to sun energy on the African continent. Unfortunately, World War I scuttled further attempts to perfect solar engineering in Africa or elsewhere. Kryza notes that solar power research "would not recover until the early 1980s."[34]

Solar Power and U.S. Economic Elites in the Post–War World II Period

While the corporate community was expressing its strong support for government (i.e., the AEC's) financial aid for the development and deployment of atomic power in the 1956 report *Peaceful Uses of Atomic Energy*, economic elites in November 1955 were noting their hostility toward government support of solar power. This opposition was outlined at the World Symposium on Applied Solar Energy. The symposium was organized by AFASE and financed in part by the Rockefeller and Ford Foundations.[35] Historian Harvey Strum explains that AFASE (founded in 1954) "initially, . . . consisted of a group of businessmen, lawyers, financiers, and educators from Arizona and California, with funds being raised in the Phoenix area." Among the founders were "Walter Bimson, chairman of the board of Valley National Bank in Phoenix." Strum notes that the "organizers of the AFASE shared . . . [a] free-enterprise approach to energy development, and they believed that 'practical utilization' of solar energy was contingent on American industry's getting involved in solar development."[36]

General Chairman of AFASE Lewis W. Douglas gave the opening remarks to the symposium. Douglas at the time was chair of the board and director of the Southern Arizona Bank and Trust Company. In his remarks, Douglas condemned the idea that "it is the responsibility of the state to distribute scarcities according to the range of priorities of purpose which the state should have the power to determine." By contrast, he spoke in positive tones about "a free and unrestrained application of scientific knowledge, functioning within the dominion of a free society, including the market place in which most economic claims are freely adjusted."[37]

Henry B. Sargent was president of AFASE. Sargent spent his entire professional career in the utility industry and was executive vice president of Central Arizona Light and Power and later president of Arizona Edison Company. By 1955, Sargent was president and director of the American and Foreign Power Company.[38] Sargent, speaking as head of AFASE, declared that the "ultimate success or failure" of solar energy "lies largely with the business man." He added that "it is he who translates technological advances into the practical accomplishments which benefit mankind and raises the standard of living and brings about a better understanding among people."[39]

The seeming result of this laissez-faire attitude toward solar power was that this form of energy received scant research support from the federal government. In the article "American Solar Energy Policy, 1952–1982," historians Harvey Strum and Fred Strum explain that "between 1952 and 1970 the National Science Foundation (NSF) conducted almost all solar energy research, averaging about $100,000 per year."[40]

AFASE changed its name in 1963 to the Solar Energy Society (SES) and went defunct in 1970 (moving from Tempe to Melbourne, Australia). Historian Harvey Strum explains that "a relatively small number of people were working in the [solar energy] field, and the general lack of interest on the part of the federal government handicapped the organization." He reports that "at the end of the 1960s, SES officials concluded that solar energy would not replace fossil fuels until a great deal of additional research had taken place and the cost of equipment had been brought down to a competitive level."[41]

Conclusion

Through the Panel on the Impact of the Peaceful Uses of Atomic Energy and its 1956 report, *Peaceful Uses of Atomic Energy*, the corporate community in the United States manifested broad support for civilian nuclear power, as well as public financing for it. In the 1950s and 1960s, the federal government (through the AEC) financially sponsored nuclear science and engineering—including the building of nuclear facilities. The AEC's annual spending on civilian nuclear power "rose from a cost of less than $20 million in 1954 to over $100 million five years later; and the scope was greatly expanded toward reactor construction and demonstration."[42] U.S. economic elites ostensibly embraced nuclear power for strategic and geopolitical reasons (e.g., the Cold War). By contrast, it was not evident that U.S. development/deployment of industrial-scale solar power could enhance its geopolitical/hegemonic position. The result is that throughout the late nineteenth and twentieth centuries, research into the collection/ use of solar power received little governmental (or institutional) support in the United States.

A key thesis of this book is that the United States is not a "normal" country when it comes to the question of energy. On the demand side, the United States fosters increased energy consumption through urban sprawl. In the next chapter, I outline how economic elites, through the 1921 President's Conference on Unemployment, successfully championed urban sprawl as a means to counter the economic downturn in the aftermath of World War I. Through the FHA, economic elites further propagated urban sprawl in an effort to address the Great Depression.

On the supply side, the United States historically drew upon its sizable fossil fuel supplies to fill the energy demand of urban sprawl. As outlined in Chapter 5, when its domestic supply proved inadequate, the United States sought to dominate the petroleum supplies of the Persian Gulf. Indicative of the fact that the United States pursued nuclear power primarily for hegemonic reasons and not to meet domestic energy demand is the fact that the United States stopped ordering new nuclear reactors after the oil shocks of the 1970s. This is in contrast to other countries, particularly Germany and France (Chapter 5), which sought to greatly expand their nuclear power capacity in response to these shocks.

Chapter 4

Urban Sprawl as Economic Stimulus

I noted/iterated at the end of the last chapter that the United States is not a "normal" or "average" country when it comes to the question of energy. On the one hand, the United States fosters increased energy use through urban sprawl. Urban sprawl creates economic demand and serves to stabilize the economy. In this chapter, I describe how and why pro–urban sprawl policies were initiated at the federal level in the 1920s. I also outline in this chapter how the federal government accelerated its urban sprawl policies in the 1930s in direct response to the Great Depression.

On the other hand, the United States developed nuclear power not as a means to attain energy autonomy/independence but as a foreign policy strategy. As outlined in the next chapter, in the aftermath of the oil shocks of the 1970s, the United States did not roll back urban sprawl (i.e., its automobile and oil dependency), but instead focused its military/diplomatic efforts on the Persian Gulf region—an area with the majority of known petroleum reserves. Significantly, the United States did not expand its nuclear energy capacity with the oil shocks, but allowed it to stagnate. In sharp contrast, the countries of Europe sought ways to reduce oil consumption in response to the oil shocks, including growing their ability to generate electricity from nuclear power.

The U.S. Federal Government and
Urban Sprawl in the 1920s

Real estate interests in United States at the turn of the twentieth century were disseminating the techniques of automobile-centered

41

urban sprawl.[1] In addition to enhancing the economic value of land on the urban periphery, automobile-centered urban sprawl expanded the market for automobiles as well as created demand for appliances and furniture to fill the relatively large homes built on the urban outskirts.[2] During the 1920s, the federal government began to promote urban sprawl as a way to stimulate the economy.

In 1921, a presidential advisory conference was convened to recommend proposals that could deal with the economic downturn, and specifically the unemployment, that followed World War I. The conference was titled the President's Conference on Unemployment and comprised an economic elite–led policy discussion group. Among the corporate elites who were conference members were the president of the U.S. Chamber of Commerce, president of the Pittsburgh Coal Company, president of the Pelham Oil & Trust Company, president of the Illinois Central Railroad Company, president of the American Steamship Owners Association, chairperson of the Bethlehem Steel Corporation (Charles M. Schwab), and president of the National Implement & Vehicle Association. The conference was presided over by Secretary of Commerce Herbert Hoover, himself a wealthy mining engineer/businessperson.[3]

One of the conference's twelve recommendations to combat unemployment was road building. The conference argued that "a congressional appropriation for roads . . . would make available a large amount of employment."[4]

The President's Conference on Unemployment formulated its recommendations through committees. It was the Committee on Public Works that developed the conference's recommendation on road building. On this committee was James Couzens, a vice president of the Ford Motor Company. Also on this committee was Evans Woollen, president of the Fletcher Joint Stock Land Bank and member of the Economic Policy Committee of the American Bankers' Association (a trade association).[5] The committee report stated that "it is the judgment of this committee that the country should put itself behind the *better roads—more work program*, insisting that it be pushed at once to the last dollar of money that is available."[6]

This road-building recommendation was consistent with the automobile industry's political agenda, which beginning in the first decade of the twentieth century promoted the reorganization of the nation's transportation infrastructure, fostering automobile dependency.

In 1903, automobile manufacturers were supporting the American Road Builders Association and the national movement to have governments at all levels pay for roads and highways that could accommodate automobiles.[7] In 1911, the American Automobile Association sponsored the first American Road Congress.[8] At this congress, Hugh Chalmers, president of the Chalmers Motor Company, conceded that "the automobile industry is, of course, in favor of good roads and would be greatly benefitted by them," but he went on to stress that "since the roads are for all the people, they should be built by all the people, or all the people should contribute to the building of them."[9] Chalmers concluded his speech by arguing:

> I believe when the people are thoroughly aroused on this question [of the quality of the U.S. road system] and realize that the benefits of [good roads] are not for one class of people alone, but for all the people alike, that they will rise up some day and demand of the national Congress, to start with, and the State assemblies, in the second place, that they cooperate to the end that we keep pace in road improvement with all other transportation improvements of this century.[10]

Another recommendation put forward by the President's Conference on Unemployment related to home building. The conference averred that "the greatest area for immediate relief of unemployment is in the construction industry." The one type of construction the conference specifically referred to was housing, regarding which the authors of the conference report claimed that "we are short more than a million homes." The conference advised "organize[d] community action . . . to the end that building may be fully resumed."[11]

The conference's Committee on Construction Industries advised that Secretary of Commerce Hoover take a leading role in an effort to promote construction (house building) nationwide:

> It is therefore recommended that Secretary Hoover . . . appoint a committee selected from the various elements interested in construction, such as financiers, labor, engineers, architects, contractors, material manufacturers, and others to be known as the Committee on Construction Development.

The conference's Committee on Construction Industries went on to suggest that its proposed Committee on Construction Development work "in cooperation with the Secretary of Commerce." The committee explicitly argued for "the prompt removal of unnecessary or inequitable [local] limitations and restrictions which have retarded real construction activity."[12]

Consistent with the President's Conference on Unemployment Committee on Construction Industries' recommendations, in 1921 Secretary of Commerce Herbert Hoover appointed an Advisory Committee on Zoning. It put out in 1924 *A Standard State Zoning Enabling Act*. Renamed the Advisory Committee on City Planning and Zoning, in 1928 it disseminated *A Standard City Planning Enabling Act*. "Together these two documents outlined the basic principles for state and local governments to follow in implementing the comprehensive urban land-use planning agenda. Many state legislatures adopted one or both of the model enabling acts almost verbatim." Marc Weiss, in his history of suburban land planning, goes on to report that "NAREB President (1922) and community builder [i.e., suburban developer] Irving B. Heitt served on both Advisory Committees, along with nine men closely associated with the newly emerging city planning profession."[13]

Historian Adam Rome describes how the Department of Commerce under Hoover promoted urban sprawl.[14] Specifically:

> Throughout the 1920s, the Commerce Department also worked with a private philanthropic organization—Better Homes for America—to promote the ideal of homeownership. The organization produced a film entitled "Home, Sweet Home" and distributed thousands of copies of the Commerce Department's *How to Own Your Own Home* booklet. By the end of the decade, Better Homes for America had over 7,000 local chapters, and each sponsored a variety of lectures and demonstrations, including construction of model homes.[15]

These Better Homes for America chapters were consistent with the President's Conference on Unemployment Committee on Construction Industries' recommendation that "in continuation of the policy of the creation of local organizations inaugurated by the Department of

Commerce, the National Federation of Construction, the U.S. Chamber of Commerce, etc. . . . the time is ripe for their more definite and extensive organization." The Committee on Construction Industries added that "such local committees as have already been organized in the country have had a profound value in readjusting the construction situation."[16]

Historian Greg Hise notes that Secretary of Commerce "Hoover served as titular chairman of the Better Homes for America movement."[17] Moreover, Marc Weiss reports that "with the accession of Herbert Hoover as secretary of commerce in 1921, NAREB became an important and highly favored trade association working closely with the Commerce Department's new Division of Building and Housing, as well as with other federal agencies."[18]

Urban Sprawl and Consumer Durables

Urban sprawl beginning in the 1920s significantly affected the U.S. manufacturing sector. This is especially evident with automobile production, in which the automobile dependency created by urban sprawl spurred automotive consumption.

Economic historian Peter Fearon notes of the other leading industrial power in the 1920s, Great Britain, that its "economy was retarded by the weight of the old staple industries such as cotton textiles, coal, shipbuilding and iron and steel. . . ." He explains that this is "in contrast to the striking advance of the consumer durables sector in America."[19] (Consumer durables are manufactured retail items expected to last at least three years.) Thus, the U.S. economy excelled in the production of such commodities as household appliances.[20] Expanding demand for consumer durables, especially automobiles, brought about the expansion and technological sophistication of the U.S. industrial sector.[21] Economic historian Alexander J. Field contends that "almost all of the [technological] foundations for [U.S.] postwar prosperity were already in place by 1941."[22]

The most prominent feature of the consumer durables–geared U.S. industrial base was automobile production. In 1920, U.S. automobile makers sold 1.9 million automobiles, and in 1929 4.4 million. American automobile manufacturers' passenger car output represented 85 percent of the global total.[23] Fearon explains that "the influence of the automobile [on the U.S. economy] was pervasive." For example, "it provided one of the chief markets for the steel industry and for the manufacturers of glass and tyres."[24] During much of the 1920s, "nearly 17 percent of the total value of fully and semi-manufactured goods was accounted for by automotive products."[25] It is statistics like these that prompt economic historian Elliot Rosen to regard the automotive industry as the "nation's principal industry" by the 1920s.[26] Economic historian Maury Klein adds that "during the 1920s the automobile industry became one of the main pillars of the American economy."[27] Another economic historian, Richard B. Du Boff, notes that "during the 1920s, the [automotive] industry became the nation's leader in manufacturing."[28]

A 1929 government report titled *Recent Economic Changes in the United States* noted the impact that urban sprawl during the 1920s was having on the U.S. economy. The report was an extension of the 1921 President's Conference on Unemployment. The *Recent Economic Changes* report was the last of three generated as a result of the conference. It was composed by the Committee on Recent Economic Changes. On the committee was Owen D. Young, chairperson and president of major appliance manufacturer General Electric;[29] John J. Raskob, chief financial officer of both General Motors and the chemical giant DuPont; as well as Daniel Willard, president of the Baltimore and Ohio Railroad.[30] As described in its report, the Committee on Recent Economic Changes "was directed to make a critical appraisal of the factors of stability and instability; in other words, to observe and to describe the American economy as a whole." The committee produced an "analysis of post-war developments in American economic life, particularly those since the recovery from the depression of 1920–21."[31]

The committee took note of the sprawling of urban zones during the 1920s: "The private automobile and bus, with improved roads, have greatly enlarged the area within which dwellings may be located, and have permitted comparatively open developments in attractive locations, to an extent that would not have been possible before the war." Moreover, "The family's enlarged radius of movement due to the automobile . . . strengthens the call toward the suburbs."[32] The authors of the committee report made the explicit point that "the extension of residential areas in and about cities, made possible by the automobile and improved streets . . . has resulted in a remarkable suburban growth of detached houses."[33] Finally, the committee observed that "the automobile has been one of the pervasive influences affecting . . . production during recent years. In addition to its direct influence on demand," the automobile "has facilitated changes in demand in many communities," and "these changes . . . have enhanced the prosperity of some groups of manufacturers."[34]

Utilizing statistical analysis, economic historian Martha L. Olney demonstrates that the dramatic increases in the consumption of durable goods (particularly of automobiles) exceeded overall increases in income during the pre–Depression Era and the post–World War II period. It is for this reason that Olney contends that the 1920s marks the beginning of the consumer durables revolution in the United States.[35]

The Federal Government's Promotion
of Urban Sprawl in the 1930s

The federal government, beginning in the mid-1930s, initiated a program to underwrite home mortgages. It did so through the FHA. The FHA's legislative authority is found in the National Housing Act of 1934. The committee that composed this act was headed by Marriner Eccles, a wealthy Utah businessperson who was an official in the Department of the Treasury. Also on this five-person committee was Albert Deane, executive "assistant to the president" of General Motors—Alfred Sloan.[36] Eccles's committee was actually a subcommittee of the President's Emergency Committee on Housing. The President's committee included W. Averell Harriman, who was asked to participate on this issue because of "his national standing as a businessman."[37] As historian Sydney Hyman explains, "When the terms of the new housing program were finally agreed to, [Harriman] was expected to 'sell' the program to . . . the business community at large."[38] Also on the President's Emergency Committee on Housing was John Fahey, chairman of the Federal Home Loan Bank Board.[39]

The presence of Sloan (through his assistant) on a presidential housing committee is noteworthy. By this time, General Motors was selling half of all automobiles in the United States. As explained above, from the first decade of the twentieth century, automobile firms were promoting a reorganization of the nation's transportation infrastructure and spurring automobile dependency.

Automobile manufacturers were not the only supporters of a national system of automobile-friendly roads and highways. Frederic

Paxson, a historian of the U.S. highway movement, notes that many early highway "proposals had money behind them, for chambers of commerce, automobile associations, and industrial organizations" contributed politically to their fruition.[40] Nevertheless, automobile firms were persistently aggressive in promoting automobile-dependent infrastructure (i.e., roads and highways).[41] In the early 1930s, for example, when cash-strapped states began using their gasoline taxes for programs other than road building, "General Motors banded two thousand groups into the National Highway Users Conference to lobby against the practice."[42] This lobbying effort yielded the Hayden-Cartwright Act of 1934, which determined that "states which diverted the [gasoline] tax to other than road use should be penalized by a reduction in their share of federal aid."[43] Stan Luger, author of *Corporate Power, American Democracy, and the Automobile Industry*, explains that at the 1939 World's Fair, General Motors "presented a model of the future based on suburbs and highways."[44] Finally, numerous automotive-related companies, among them General Motors, Standard Oil of California, and Firestone Tire and Rubber, were found by a federal grand jury to have successfully conspired to dismantle electric streetcar (trolley) systems in forty-five U.S. cities, including Los Angeles, San Francisco, and New York, during the 1940s.[45]

Marriner Eccles's biographer (drawing from extensive interviews with Eccles) outlines the thinking underlying the formulation of the National Housing Act of 1934. "A program of new home construction, launched on an adequate scale, would not only gradually provide employment for building trade workers, but"—more importantly— "accelerate the forward movement of the economy as a whole." It was anticipated that

> Its benefits would extend to everyone, from the manufacturers of lace curtains to the manufacturers of lumber, bricks, furniture, cement and electrical appliances. Transportation of supplies would stimulate railroad activity, while the needs generated for the steel rails and rolling stock would have spin-off effects on steel mills.[46]

Moreover, "if banks with excess reserves made loans for home construction, the effect would be to create the basis for new money."[47]

Therefore, the purposes of the legislation that authorized the FHA were seemingly to spur consumption, including that of consumer durables (retail items expected to last at least three years [e.g., automobiles]), and to prompt the profitable movement of capital out of banks and into the housing sector. Urban sprawl would presumably help to accomplish these goals, because suburban developers by the 1920s already demonstrated a predilection for building large, relatively expensive homes on undeveloped tracts of land, far from trolley lines.[48]

Upon its creation, the FHA was placed under the stewardship of prominent officials from the real estate sector, and they used their authority to promote the horizontal growth of urban American. Created in 1934:

> FHA's staff was recruited almost entirely from the private sector. Many were corporate executives from a variety of different fields, but real estate and financial backgrounds predominated. For example, Ayers DuBois, who had been a state director of the California Real Estate Association, was an assistant director of FHA's Underwriting Division. Fred Marlow, a well-known Los Angeles subdivider, headed FHA's southern California district office, which led the nation in insuring home mortgages. National figures associated with NAREB, such as real estate economist Ernest Fisher and appraiser Frederick Babcock, directed FHA operations in economics and in underwriting.[49] [Significant for this discussion is the fact that the first administrator of the FHA was an executive from Standard Oil. Also among the FHA's initial leaders were two individuals from the automotive sector: Albert Deane of General Motors, who was deputy administrator of the FHA, and Ward Canaday, "president of the U.S. Advertising Corporation of Toledo, with a reputation for sales promotion in the automobile industry." Canady was the FHA's director of public relations.[50]]

Jeffrey Hornstein, a historian of the U.S. real estate industry, notes that the industry generally "welcomed the FHA . . . both because it promised greatly enhanced general demand for housing and because the agency was run largely by Realtors and their allies in the banking world."[51]

As a way to encourage housing sales, the FHA underwrote home purchases. It would guarantee 80 percent of home mortgages for qualified homes and buyers for a twenty-year term. (Later, this guarantee was modified to 90 percent and twenty-five years.[52]) Up to this time, standard mortgages covered about 50 percent of the home purchase price and had a three-year term.[53]

This program gave the FHA the ability to influence the types of homes purchased and, subsequently, housing development patterns. Weiss notes:

> Because FHA could refuse to insure mortgages on properties due to their location in neighborhoods that were too poorly planned or unprotected and therefore too "high-risk," it definitely behooved most reputable subdividers to conform to FHA standards. This put FHA officials in the enviable position, far more than any regulatory planning agency, of being able to tell subdividers how to develop their land.[54]

With this power, the FHA promoted the building of large-scale housing developments in outlying areas. Weiss explains that the Federal Housing "Administration's clear preference . . . was to use conditional commitments [for loan guarantees] specifically to encourage large-scale producers of complete new residential subdivisions, or 'neighborhood units.'" Thus, the FHA, through its loan program, encouraged and subsidized "privately controlled and coordinated development of whole residential communities of predominately single-family housing on the urban periphery."[55]

With federal housing policy firmly under the control of the FHA, Kay writes that it "decentralized housing out of the city and did little to help slum dwellers."[56] In his comprehensive analysis of U.S. suburban development, geographer Peter O. Muller explains that "the nearly complete suburbanization of the [urban middle class] . . . was greatly accelerated by government policies . . . the most important being the home loan insurance programs launched by the Federal Housing Administration in 1934."[57] Kay adds:

Cities remained the center of Depression malaise and neglect. Their expansion ceased or declined compared to suburbs. Twenty-five percent of Detroit's growth was on its periphery, only 3 percent within the city. Likewise, Chicago's suburbs swelled 11 percent, the downtown less than 1 percent. Vast acreage in the central business districts fell for parking spaces.[58]

Kenneth Jackson, in his important history on the suburbanization of urban development in the United States, concurs with Weiss's, Kay's, and Muller's assessments of the bias within the FHA for new housing stock in outlying areas. Jackson writes that "in practice, FHA insurance went to new residential developments on the edges of metropolitan areas, to the neglect of core cities."[59] As a result, Jackson notes, between the years 1942 and 1968, the "FHA had a vast influence on the suburbanization of the United States."[60]

As noted above, urban sprawl in the United States spawned a consumer durables revolution beginning in the 1920s. This revolution was sustained throughout the postwar period, with the consumption of consumer durables (especially of automobiles) substantially exceeding income growth.[61] Today, U.S. urban sprawl has international economic ramifications. The United States is the world's largest consumer.[62] (U.S. consumers, excluding government and businesses, purchase close to 20 percent of the world's total economic output.[63]) Importantly, European, Japanese, and South Korean automakers count heavily on access to the huge U.S. automobile market to attain profitability.[64] With a population one-third smaller than those of Western and Central Europe, before the recent recession, the United States consumed on average 2 million more automobiles annually (during peak years: 15 million versus 17 million), and, at least until recently, half of all new automobiles purchased in the United States were of the highly profitable SUV and light truck varieties.[65] The Japanese automakers Honda and Toyota (the world's largest automobile manufacturer), for instance, in the near past derived two-thirds of their overall profits from sales in the United States.[66] It is also noteworthy that General Motors, the second largest manufacturer of automobiles, registers about 40 percent of its sales in the U.S. market.[67] (The United States contains less than 5 percent of the world's population.)

Conclusion

I started this chapter by stressing that the United States is not a "normal" country when it comes to energy. Policy makers foster profligate use of domestic stocks of fossil fuels to increase demand for consumer durables through urban sprawl. By the 1930s, the federal government was using mortgage subsidies offered through the FHA to accelerate urban sprawl. Today, policy makers promote urban sprawl with cheap energy policies (e.g., low gasoline taxes and occupying oil-rich countries [e.g., Iraq]), liberal land policies, and road building into sparsely populated areas.

In the 1970s, it became evident that the United States was dependent on the volatile global oil market. It did not seek to insulate itself from this market through conservation measures (i.e., limiting automobile use), but predominantly through the projection of military/political power. The countries of Western and Central Europe did seek to limit their exposure to the world oil market by reducing petroleum consumption. Part of this strategy involved expanding their nuclear capacity. Somewhat counterintuitively, the United States turned away from civilian nuclear power in the aftermath of the oil shocks. This serves to support the argument that the United States' nuclear capacity was intended as a foreign policy device and not as a means of ensuring domestic supplies of energy.

Chapter 5

Global Oil Politics

As I explained in Chapter 1, Frank N. Laird, in his book *Solar Energy, Technology Policy, and Institutional Values*, holds that the U.S. aggressively deployed civilian nuclear power and forewent solar energy because of the ideas that dominated White House thinking on energy. Laird explains that during the post–World War II period, the ideas of energy supply and national security were linked, and beginning with the Eisenhower administration (1953–1961), these concepts became tied to nuclear power—which held the promise of virtually limitless and inexpensive supplies of energy.[1]

U.S. oil policy indicates, however, that domestic energy supply and demand issues have never predominated the American government. Instead, the idea that has driven U.S. energy policy has been to sustain urban sprawl in America (at virtually all costs). This becomes particularly evident in the aftermath of the oil shocks of the 1970s, when the U.S. government, instead of curbing demand (i.e., urban sprawl), sought to dominate the petroleum supplies of the Middle East. The U.S. also began to wind down its civilian nuclear power capacity after the oil shocks.[2]

By contrast, the countries of Western and Central Europe have tried to bolster their energy security by limiting their exposure to the international energy market. First, by limiting automobile dependency, and second, by expanding domestic nuclear power in the aftermath of the 1970s oil shocks.

U.S. Oil Policy

In 1973, the Persian Gulf region of the Middle East took on particular importance for the Western allies. What came into relief in
1973 is that the region contained the key supplies of petroleum for
the Western world. The petroleum-bearing countries of the region are
Iran, Iraq, Kuwait, Saudi Arabia, United Arab Emirates, and Qatar,
with Iran, Iraq, Kuwait, and Saudi Arabia being the primary producing
countries for the world's oil market. The Persian Gulf nations today
possess the majority of the world's known petroleum reserves—Saudi
Arabia alone is estimated to hold 20 to 25 percent of the world's
proven reserves of petroleum.[3]

The Persian Gulf's strategic importance is in significant part
the result of U.S. oil policies. This is particularly apparent on the
demand side. As U.S. cities became more and more sprawled[4] and as
a result more automobile dependent,[5] U.S. oil consumption steadily
climbed.[6] Between 1946 and 1953, for instance, U.S. gasoline usage
went from 30 billion gallons annually to 49 billion, amounting to a
yearly growth rate of slightly more than 7.2 percent. In 1958, U.S.
gasoline consumption exceeded 59 billion gallons.[7]

U.S. consumption had a detrimental affect on its petroleum
production. This was important because the United States was historically capable of reducing world petroleum prices through increased
production. By 1970, however, U.S. oil production had peaked, and
it was no longer capable of regulating world prices.[8] When Saudi
Arabia imposed a selective embargo on countries favorable to Israel
in 1973, the United States was importing close to 40 percent of its
oil needs, and it could not respond to the shortfall created by the
embargo with domestic production.[9]

The Oil Shocks of the 1970s

Therefore, leading up to the oil shocks of the 1970s, U.S. oil reserves
were depleted predominately because of high levels of domestic consumption. What is theoretically and historically significant, however,
is the response of the U.S. government when the vulnerability of the
U.S. economy and its dependency on foreign sources of petroleum
came into stark relief in 1973. No effort was put forward by the U.S.

government to roll back or limit urban sprawl and the automobile dependence that it spawned.

The United States responded militarily to its apparent dependency. U.S. policy makers used the country's superior political and military position to ensure that Persian Gulf oil remained in the U.S. sphere of influence and that the region's petroleum sufficiently flowed. Until 1979, the United States amply supplied the Iranian government with military equipment and training to militarily safeguard the petroleum reserves of the region against any Soviet aggression. After the United States' client regime in Iran collapsed (which brought on a second oil crisis), the United States sought to directly build up its military capabilities in the region, culminating with a direct military presence after the first Persian Gulf War in 1991.[10]

This emphasis on the supply side to deal with the United States' energy problems of the 1970s is reflected in two reports put out by the Twentieth Century Fund (now the Century Fund). This organization is a foundation that in the 1950s and 1960s sponsored studies on the natural resource needs of the United States' expanding economy.[11] The Twentieth Century Fund created two policy-planning groups in the early 1970s composed largely of economic elites who put forward proposals to deal with the United States' petroleum situation. One task force, convened in 1973, was named the Twentieth Century Fund Task Force on United States Energy Policy. On this task force were a director and senior vice president of Exxon, a vice chairman of the board of the American Electric Power Company, Walter J. Levy (a consultant to most major oil firms),[12] a vice chairman of the board of Texas Commerce Bancshares (a major Texas bank),[13] and the chairman of the board of Carbomin International Corporation (an international mining firm). The other task force, formed in 1974, was known as the Twentieth Century Fund Task Force on the International Oil Crisis. Walter J. Levy and the executives from Carbomin and Texas Commerce Bancshares also served on this task force. Also on this Twentieth Century Fund task force were the chairman of the board from Atlantic Richfield (an oil firm), a managing director from Dillon, Read & Co. (a leading New York investment management firm), the chairman of the board from the Louis Dreyfus Corporation (an investment management firm), the chairman and president of The First National Bank of Chicago, and a consultant to Wells Fargo

Bank (a major California bank). The task forces were also composed of academics (mostly economists) from Princeton, Harvard, MIT, and the University of Virginia, as well as the presidents of Resources for the Future (which was on the two task forces) and the Carnegie Institution (only on the energy policy group)—both of which are economic elite–led research institutes.[14]

In the wake of the 1973 oil shortage and the Organization of Petroleum Exporting Countries (OPEC) seeking to maintain high oil prices, both of the Twentieth Century Fund's task forces advised that the United States should strive to develop sources of oil and energy outside the OPEC countries. This would serve to reduce the strategic positioning of OPEC countries over petroleum and petroleum prices. OPEC includes all of the Persian Gulf oil producers plus Algeria, Angola, Ecuador, Libya, Nigeria, Venezuela, and (until 2009) Indonesia. The Twentieth Century Fund's Task Force on the International Oil Crisis advised that "the best remedy for the problems caused by the increased price of oil [brought about by OPEC members] would be, simply, to lower the price" of petroleum. "The Task Force believes that this remedy should be sought through reliance on market forces."[15] The task force goes on to explain in its report that "the most effective means of exerting market pressure will be to accelerate exploration for crude and develop producing capacity from" areas outside of OPEC.[16] The task force on U.S. energy policy averred "that it is essential that the nation take firm and forceful action to implement a comprehensive near-term energy program designed to assure greater availability of domestic supplies of oil and other sources of energy."[17] The authors of this task force's report went on to explain:

> Our present dependence on OPEC cannot be eliminated, but it can—and should—be lessened, thus reducing the competition for OPEC supplies and consequently the political and economic power of the cartel. While we cannot achieve independence, a lessening of our dependence can make a disruption of supplies or a more aggressive price policy on the part of OPEC much less likely.[18]

Therefore, the key recommendations put forward by these policy-planning groups, made up in large part of economic elites, in light of U.S. oil dependency on OPEC countries were to expand the supply of available energy free from OPEC control and not necessarily to reduce energy consumption.

Both of these groups, in their reports, called for greater energy efficiency, or what they labeled in their reports as "conservation." The difficulty is that increased energy efficiency does not necessarily reduce overall consumption levels. The energy policy group, in a section of its report titled "Measures to Promote Conservation," "*endorse[d] the use of special incentives to encourage further investment in energy-saving capital goods and consumer durables because conserving energy is as important as increasing the supply.*"[19] It specifically suggested in its report the use of a "luxury" tax to discourage the purchase of large, less efficient automobiles. Moreover, the implementation of "excise taxes levied annually and collected with state registration fees also might serve to encourage quicker scrapping of cars that consume above-average amounts of gasoline."[20] Finally:

> The Task Force favor[ed] the continuation of such energy-conserving measures as reasonable speed limits on highways, building standards that reduce the use of energy for heating and cooling, and requirements that appliances bear tags disclosing their energy-utilization efficiency.[21]

The Task Force on the International Oil Crisis did not set out specific conservation proposals. Instead, it deferred to the energy policy task force on this.[22]

Increased energy efficiency can lead to overall lower levels of petroleum consumption. Energy savings from increased efficiency, however, can be offset by increased economic growth.[23] This is especially the case within sprawled urban regions, where greater levels of economic activity can lead to a larger workforce driving to and from work and increased demand for spacious homes on the urban periphery. So whereas automobiles may become more fuel efficient, in the context of diffusely organized cities, more automobiles and longer driving distances can lead to greater overall gasoline/oil consumption—in spite of gains made in fuel efficiency.[24] This is precisely what has transpired in the United States. The current U.S. automobile fleet is more efficient than the U.S. automotive fleet of the early 1970s.[25] Because, however, of a substantially enlarged automobile population and ever-increasing amounts of driving, gasoline/diesel consumption in the United States

today substantially exceeds that of the 1970s. According to energy
economist Ian Rutledge, in 1970, driving in the United States consumed
7.1 million barrels per day of petroleum, whereas by 2001 that figure
had increased to 10.1 million.[26] Today, according to the U.S. Department of Energy, automobile driving in the United States consumes
more than 10 percent of total global oil production.[27]

Because in large part of the steady growth of gasoline/diesel
consumption in the United States,[28] its economy consumes about 25
percent of the world's total petroleum production (with less than 5
percent of the global population).[29] This is especially glaring because
in the aftermath of the spike in oil prices in the 1970s, U.S. factories
and utilities shifted from petroleum-based fuels to other sources of
energy (mostly coal, natural gas, and nuclear power).[30]

It is noteworthy that the Twentieth Century Fund Task Force
on United States Energy Policy in 1977 "recommend[ed] an extensive
program of government-supported research and development for new
energy sources." The task force specifically pointed to oil shale or
synthetic gas derived from coal as potential alternatives to petroleum-
based gasoline. It also advised government funding "to develop the
more exotic alternative energy sources."[31] In response to the oil shocks
of the 1970s, the federal government did invest significant sums of
money in investigating alternative fuels, but as petroleum prices fell
by the early 1980s, it dramatically cut this spending.[32]

It is telling that neither of the Twentieth Century Fund's task
forces counseled less driving or mass transportation as conservation
measures to counter OPEC price strategies. Such a recommendation
would have raised urban sprawl and the automobile dependence that
it creates as political issues.

European Post–World War II Oil Policies

The advanced industrialized countries of the Federal German Republic
(i.e., West Germany) and France responded to the oil crises of the
1970s by trying to severely limit their use of it. These nations had
little appreciable domestic petroleum production. Because of major
oil strikes in 1966 along its northern coast, Great Britain had less
immediate need to reduce its oil use. Nevertheless, the nations of

Western Europe had not developed the petroleum vulnerability that the United States had by the 1970s. This is particularly the case because urban zones in these countries were not as sprawled and automotive dependent as those in the United States.[33]

Postwar Western European concerns about energy security were manifest in the 1955 Armand report and the 1956 Hartley report, both sponsored by the Council of Ministers of the Organization for European Economic Cooperation (OEEC). Primarily because of fear over trade imbalances, the Armand report, titled *Some Aspects of the European Energy Problem: Suggestions for Collective Action* and named after its author, Louis Armand (a French government official), advised against dependency on foreign sources of oil. Instead, Armand advised Western European countries to rely on domestic sources of energy, on sources of energy that were in Europe's African colonies, and especially on nuclear power.[34]

Shortly after receiving the Armand report, the OEEC created a Commission for Energy. The commission sponsored what became known as the Hartley report, named after its chairperson, Harold Hartley of Great Britain. The authors of this report extended their concerns over oil imports beyond trade imbalance issues and expressed fears about oil security. According to the Hartley report, "there are inevitable risks in the increasing dependence on Western Europe on outside [oil] supplies, particularly when most of them must come from one small area of the world" (i.e., the Persian Gulf).[35] Accordingly, Western Europe by 1975 should only draw 20 to 33 percent of its energy from imported petroleum, and the rest should come predominately from coal.[36] The Hartley report authors averred that "coal must remain the mainstay of the Western European energy economy."[37] They recommended that Western European domestic coal production satisfy half of the region's energy needs, and the rest could be met with hydropower, natural gas, oil, imported coal, and nuclear energy.[38]

Both the Hartley and Armand reports counseled that Western European governments should intervene to ensure the region's energy stability. The Hartley commission suggested that

in order to deal effectively with the urgent problems involved in the supply and demand of energy, each Member country will require an energy policy suited to its own circumstances and its needs and resources. This policy should include some measure of coordination between the different forms of energy.[39]

Armand held that OEEC countries should avoid "a situation in which competition between the various forms of energy acts to the detriment of the community as a whole."[40]

Subsequent to the Armand and Hartley reports, the OEEC formed the Energy Advisory Commission under the chairperson-ship of Professor Austin Robinson. In 1960, this energy commission put forward a new report on European energy titled *Towards a New Energy Pattern in Europe*. Unlike the Armand or Hartley reports, which advocated government promotion of domestic coal (Hartley) or nuclear power (Armand) in order to limit imported oil use, the Robinson commission argued that Western Europe should rely on inexpensive imported petroleum for much of its energy needs. As to the security of oil supplies, new discoveries in Venezuela, West Africa, and Libya and, "in particular, discoveries of oil and natural gas in the Sahara [e.g., Algeria] have created new possibilities of important sup-plies in an area more closely integrated into the economy of Western Europe." Therefore, "as a result, there has been made possible a wider diversification of oil supplies to Western Europe."[41] The Robinson energy commission went on to argue that "it does not seem likely that shortages of oil or other supplies will make themselves felt in acute form by 1975."[42] With regard to the region's balance of payments, the commission asserted that "if Western Europe can maintain its share of world markets for manufactures, the import of the increased pro-portions of the total supplies of energy that have emerged from our study may reasonably be expected to be within the probable limits of its capacity."[43] Hence, the way to cover the costs of imported energy is to maintain or expand Western Europe's world market share of industrial products. A key means to do this is to keep the cost of energy inputs low. Thus, "when formulating a long-term energy policy, the paramount consideration should be a plentiful supply of low-cost energy. . . ." Additionally, "the general interest is best served by placing the least possible obstacles in the way of economic development of the newer and cheaper sources of energy."[44] In other words, Western

European governments should not subsidize nuclear power or coal to the detriment of abundant and inexpensive petroleum supplies.

Especially in the areas of electricity and industrial production, as well as home heating, Western European countries did pursue the more liberal course advocated by the Robinson commission. As a result, by the early 1970s, 60 percent of this region's energy needs were met through imported oil.[45] In the case of automobile transportation, however, Western European countries have historically instituted more restrictive policies. Haugland and his associates, experts on European energy, point out that in Western and Central Europe, "the share of taxes in transport fuel—in particular for gasoline—is generally the highest of all end-use prices. In Europe the tax share in unleaded gasoline [for example] is substantially above the actual production costs, ranging from 50 to 75 percent of the end-user price." They go on to assert that "not surprisingly, in the United States, where gasoline taxes are the lowest in the OECD [Organization for Economic Cooperation and Development], the average fuel consumption ranks among the highest."[46] By way of comparison, whereas the average cost of gasoline was recently $2.68 per gallon in the United States, it was $7.00 in Britain, $7.19 in Germany, $6.97 in Italy, and $6.89 in France. These price differences are mostly, if not solely, attributable to taxation.[47] On a per capita basis, the United States uses more than twice as much gasoline as these other countries.[48]

There is a strategic advantage to limiting oil use in the realm of transportation but allowing it to expand in such areas of the economy as electricity and industrial production. There are readily available substitutes for petroleum products in these latter activities: coal, natural gas, nuclear power, wind power, solar, and so forth.[49] This is not the case for automotive transportation. Thus, if there is a severe shortage of crude, the housing stock, industrial infrastructure, and retail outlets that are only accessible via automobile can become virtually worthless overnight.

With the oil shortages of the 1970s, the governments of France and West Germany sought to slash their petroleum consumption by greatly expanding the use of nuclear power. This strategy, however, sparked the green environmental movement on the continent,[50] as the question of what to do with the highly radioactive waste from nuclear power production has never been satisfactorily answered.[51]

This movement was more successful in Germany than in France in derailing plans to center industrial and electricity production on nuclear energy. Political scientist Michael Hatch contends that these different outcomes can be attributed to each country's respective political system. The French employ a presidential system in which policy-making power is in large part insulated from the public in the executive branch. The German parliamentary form of government is more sensitive and responsive to social movements and strong shifts in public opinion.[52] Nevertheless, France's shift to nuclear power,[53] the more modest increase of nuclear power in other countries of the region,[54] greater use of coal and natural gas, and increases in energy efficiency did result in a decline in petroleum consumption in Western Europe, whereas oil consumption in the United States increased after the energy shocks of the 1970s.[55]

It is worthwhile to outline France's move toward a nuclear-centered economy in the mid-1970s. (Recall that virtually at the same time the United States stopped ordering new nuclear reactors.) In 1974, the French government announced a plan to expand its nuclear program, projecting thirteen nuclear plants of 1,000 Mw each to be completed by 1980. The long-term plan was to build by 1985 fifty reactors in twenty locations providing 25 percent of France's energy and by the year 2000 two hundred reactors in forty nuclear parks providing more than half of France's projected energy needs. By the 1980s, France became more reliant on nuclear power than any other country, drawing upward of 75 percent of its electricity from this source. Today, fifty-nine nuclear power reactors operate in France. It is the second largest producer of electricity from nuclear energy and the largest exporter of electricity in the world.[56]

Conclusion

The United States undermined its national energy security by promoting, fostering, and maintaining sprawled urban zones (as described in the last chapter). Even when it was evident that the United States was dependent on foreign petroleum to meet is energy needs (i.e., the oil shocks of the 1970s), no effort was undertaken to curb the key source of its energy vulnerability—urban sprawl. Instead, the United States responded by seeking to dominate those regions of the world where surplus petroleum is located—most glaringly the Persian Gulf area.

The behavior of Western and Central Europe is consistent with the traditional or "normal" notion of national energy security. The countries of this region have tried to limit their exposure to the world energy market by limiting automobile dependency. After the oil shocks of the 1970s, France in particular significantly expanded its nuclear capacity as a means of bolstering its energy security and economic stability.

Chapter 6

Global Energy Politics and Urban Sprawl

As described in the last chapter, more than 30 years ago, the Twentieth Fund Task Force committees on oil and energy advised that the prime strategy the United States should pursue with regard to energy, especially oil, was to bring more onto the global market. American foreign policy over the last thirty-plus years has remained focused on ensuring that the price of petroleum remains relatively low by ensuring an ample supply of oil on the world market. This has been accomplished by concentrating U.S. diplomatic and military attention on the region of the world with the lion's share of proven petroleum reserves—that is, the Persian Gulf. America's global oil strategy included repulsing the Iraq invasion of Kuwait in the early 1990s (i.e., the first Persian Gulf War).[1]

Although U.S. policy has been successful in attaining American hegemony over the Persian Gulf region, the question of world petroleum supply nonetheless remains. Obviously, oil is a finite resource. The other energy-related issue that has come to the international forefront is global warming. Global policy groups have responded to concerns about petroleum supply and global warming by arguing for the development of technologies (including alternative fuels). As noted in the preceding chapter, in the mid- and late 1970s, the Twentieth Century Fund Task Force committees on oil and energy advised research into alternative fuels and greater energy efficiency. Today, the Asia-Pacific Partnership for Clean Development and Climate, the US-China Clean Energy Forum, the WBCSD, and the ICC argue that the appropriate response to issues of fossil fuel depletion and climate change is technological efficiency and alternative energy (i.e., the "weak" ecological modernization of the economy).

The fact that the U.S. government until recently has virtually neglected solar energy makes a conversion to a "clean" alternative energy infrastructure in the near future much less probable. A recent RAND report, for instance, concluded that much more research is required to determine the commercial and military viability of alternative liquid fuels (especially those derived from solar power).[2] In a 2011 report titled *Special Report on Renewable Energy Sources and Climate Change Mitigation*, the United Nations' Intergovernmental Panel on Climate Change (IPCC) holds that there is great potential in solar energy, but significant monies have to be invested to bring solar power (and other non-nuclear) alternatives to fruition.[3] Therefore, the American effort to address greenhouse gas emissions and fossil fuel depletion through alternative energy is greatly handicapped by the fact that the U.S. government has not historically made significant investments into solar energy.

The Project for a New American Century, another economic elite–led policy discussion group, posited a more sinister solution to the political/economic threat posed to the United States by declining global petroleum supplies—the conquest of Iraq (the country with reputably the second largest proven reserves of oil[4]). Dick Cheney, a wealthy oil executive, was a leader of the Project for a New American Century and later, as vice president under the administration of George W. Bush, successfully championed the 2003 Iraq invasion. Prior to the invasion, Cheney headed in early 2001 the National Energy Policy Development Group, a government group that argued that increasing U.S. oil demand required increasing access to petroleum supplies.

If U.S. energy demand does continue to grow, urban sprawl will be a prime cause. Therefore, urban sprawl, and the oil dependency it spawns, create a significant vulnerability for the United States in a context where global petroleum production is seemingly stagnating and potentially on the cusp of a decline.

Urban Sprawl and Global Oil Production

Urban sprawl is predicated on abundant global oil supplies. As many major cities worldwide expand horizontally, petroleum in excess of demand is necessary to maintain gasoline/diesel inexpensive enough to make daily commuting, as well as numerous other automotive trips, economically feasible. With oil surpluses available, extraction and production can readily rise as demand for gasoline/diesel increases, thus keeping the price of petroleum stable and relatively low.

Given the degree of urban sprawl of U.S. cities in particular and their subsequent automotive and oil dependency, the potential for the disappearance of surplus oil bodes disaster for the U.S. economy, and by implication the global economy. What will occur with the exhaustion of excess global petroleum is that oil supplies will become tighter as demand stays the same or increases, hence driving up prices to potentially damaging levels. The worst scenario is one in which global petroleum production peaks and subsequently declines and the demand for petroleum products remains stable or climbs. This would ultimately create an economically destabilizing gap between supply and demand. Examples of such a gap occurred in the United States in the 1970s, when for short periods of time oil supply fell substantially short of demand—causing the price of gasoline to reach meteoric highs, its rationing through price controls, and long lines to purchase it. In both cases, the shortfall of petroleum triggered severe global recessions.[5]

The type of petroleum production peak that eventually could wreak havoc on the U.S. economy is known as Hubbert's peak, named after M. King Hubbert—a long-time geologist for Shell. Comparing rates of petroleum extraction with the amounts of newly discovered oil, Hubbert in 1956 famously predicted that U.S. crude production would peak in the early 1970s. It did so in 1970, at somewhat less than 10 million barrels a day.[6] Hence, when the U.S. economy was suddenly cut off from Saudi Arabian oil in 1973 and from Iranian supplies in 1979, U.S. production could not exceed (or meet) the 1970 peak. (Today, in spite of significantly increased drilling activity including hydraulic fracturing, petroleum production in the United States is at about 8 million barrels a day).[7]

Expert observers have concluded that global peak petroleum production is set to occur in the near future, if it has not already occurred.[8] Energy expert Roy L. Nersesian observed in 2007 that "the frequency of discovering major oil fields is dropping; the size of newly discovered oil fields is falling; and consumption is getting ahead of additions to proven reserves."[9] Following from Hubbert's theory of oil extraction, these observations portend global peak petroleum production. Petroleum geologist and former colleague of Hubbert (who died in 1989) Kenneth Deffeyes holds that the peak occurred in 2005.[10] The International Energy Agency, a public body that advises twenty-eight industrialized countries (including the United States) on energy matters, holds that global peak oil production occurred in 2006.[11] R. W. Bentley, from Great Britain's Reading University and the Oil Depletion Analysis Center in London, pushes back the point of peak production to between 2010 and 2015.[12]

There are indications that peak production is occurring sooner rather than later. In July 2008, the cost of a barrel of oil surged to $147, representing a more than sevenfold increase since 2002.[13] (During the oil crisis of the 1970s, the price of a barrel of oil [adjusting for inflation] peaked at just over $100.) Paul Roberts, writing in early 2005, reports:

> At an energy conference in Houston last spring, Saudi oil officials admitted that production at their largest fields was being maintained only by the injection of massive volumes of seawater to force the oil to keep flowing out. They also admitted that Ghawar, the largest oil field ever discovered and a mainstay of the world oil business, was more than half depleted and that the reserves in parts of Ghawar were down to just 40 percent of their original volume.

Roberts adds:

> At the same conference, Matt Simmons, a Houston energy investor and Bush administration energy adviser who has studied trends in world oil production, made the case that Ghawar is actually closer to 90 percent depleted and that the Saudi oil kingdom is much nearer its production peak than anyone in Riyadh—or Washington—wants to believe.[14]

Simmons, in his book about Saudi oil production titled *Twilight in the Desert: The Coming Saudi Oil Shock and the World Economy*, makes the specific claim that "Saudi Arabian [petroleum] production is at or very near its *peak sustainable volume* (if it did not, in fact, peak almost 25 years ago), and is likely to go *into decline* in the very foreseeable future."[15]

The U.S. Invasion of Iraq

With Saudi oil production seemingly waning and in danger of faltering, Iraqi oil takes on great strategic importance. This is for two reasons. First, freed from the sanctions imposed on it in the aftermath of the first Persian Gulf war, Iraqi oil reserves (reputed to be the second largest in the world) could serve to ease otherwise tightening world oil supplies and could push any potential peak in global oil production into the future. Second, with major oil-producing countries declining in rates of extraction (e.g., the United Kingdom, and Indonesia),[16] Iraq's oil reserves, many of them untapped, rise in geopolitical value.[17]

Viewed through the prism of U.S. urban sprawl, U.S. policy toward Iraq is not only intended to gain greater geopolitical hegemony[18] or to enhance the profit potential of U.S. oil firms,[19] but also to facilitate the persistence of U.S. urban sprawl. The additional oil potentially on the world market from the Iraq invasion would allow urban sprawl to persist and expand.[20] Moreover, because of America's dependency on petroleum due to urban sprawl, the United States would be particularly vulnerable to a hostile Iraq regime—or even one inclined to maximize rent taking because of the increasingly key market position of its oil reserves. The invasion, which took place in March 2003, was spearheaded by a policy network composed of individuals widely known as "neoconservatives."[21] Before I outline the neoconservative "takeover" of the U.S. foreign policy/military apparatus, I turn to the National Energy Policy Development Group, in which Dick Cheney, a neoconversative leader, helped outline why the United States needed Iraqi oil to fully reenter the global market.

National Energy Policy Development Group

As vice president, Dick Cheney officially headed the National Energy Policy Development Group. This group was established by President George W. Bush during the second week of his presidency.[22] The energy policy group released its report, *National Energy Policy*, in May 2001. Officially composed of cabinet secretaries (including state, treasury, and energy) as well as top government administrators (e.g, director of the Environmental Protection Agency [EPA] and the deputy chief of staff for policy), the National Energy Policy Development Group conferred with representatives from a number of major U.S. energy-related corporations and trade groups as well as business lobbyists. According to a General Accounting (now Accountability) Office report:

> Senior agency officials [as part of the National Energy Policy Development Group] participated in numerous meetings with nonfederal energy stakeholders to discuss the national energy policy . . . including the American Coal Company, Small Refiners Association, the Coal Council, CSX, Enviropower, Inc., Detroit Edison, Duke Energy, the Edison Electric Institute [utility firm trade association], General Motors, the National Petroleum Council [oil trade association], and the lobbying firm of Barbour, Griffith & Rogers.
>
> [R]ecommendations, views, or points of clarification [were solicited] from other parties. For example, . . . detailed energy policy recommendations [were solicited] from a variety of nonfederal energy stakeholders, including the American Petroleum Institute [petroleum trade association], the National Petrochemical and Refiners' Association, the American Council for an Energy-Efficient Economy, and Southern Company. [R]ecommendations [were received] from others, including the American Gas Association, Green Mountain Energy, the National Mining Association, and the lobbying firms the Dutko Group and the Duberstein Group. Senior EPA officials, in addition to accompanying the Administrator to meetings with nonfederal energy stakeholders, discussed issues related to the development of an energy policy at meetings with the Alliance of Automobile Manufacturers, the American Public Power Association, and the Yakama Nation Electric Utility.[23]

The National Energy Policy Development Group, in formulating its energy proposals, starts by writing that "America's energy challenge begins with our expanding economy."[24] More specifically, under a figure graphing the expanding gap between U.S. petroleum consumption and domestic production, the Cheney energy group concluded that:

> Over the next 20 years, U.S. oil consumption will grow by over 6 million barrels. If U.S. oil production follows the same historical pattern of the last 10 years, it will decline

by 1.5 million barrels per day. To meet U.S. oil demand, oil and product imports would have to grow by a combined 7.5 million barrels per day. In 2020, U.S. oil production would supply less than 30 percent of U.S. oil needs.[25]

While pointing to efficiency as a conservation strategy, the National Energy Development Group stressed meeting increased energy demand through increased supply: "A primary goal of the National Energy Policy is to add supply from diverse sources. This means domestic oil, gas, and coal. It also means hydropower and nuclear power. And it means making greater use of non-hydro renewable sources now available."[26]

A seemingly more realistic way to address the energy group's predicted sharp increase in U.S. oil demand is for the United States to take direct control of Iraq's oil fields, which had been significantly underutilized since the 1990 Persian Gulf War. Although the Cheney energy group emphasized domestic sources of energy, its report did include a chapter titled "Strengthening Global Alliances: Enhancing National Energy Security and International Relationships," in which it was explained that "Middle East oil producers will remain central to world oil security," and the Persian "Gulf will be a primary focus of U.S. international energy policy."[27] The public record shows that the Cheney energy group actively discussed Iraqi oil reserves.[28] The energy group noted in its report that "by 2020, Gulf oil producers are projected to supply between 54 and 67 percent of the world's oil. . . . *This region will remain vital to U.S. interests.*"[29] In 2003, Dick Cheney, as leader of the neoconservatives in the George W. Bush administration, brought about the U.S. invasion of Iraq.

The Neoconservative Policy Network and Iraq

Neoconservative pundits date back to the early Cold War period. These were a group of thinkers who embraced domestic New Deal programs but rejected efforts on the left to establish a congenial foreign policy toward the Soviet Union. Neoconservatives argued for an aggressive, confrontational tack toward the Soviet Union and often derided realist rapprochement with the Soviets and even containment. Another hallmark of neoconservative thinking is its strongly pro-Israel bent. Therefore, throughout most of the postwar period, the neoconservative grouping formed part of the so-called "right" within the Democratic

Party. In the 1980s, prominent neoconservatives joined the Reagan administration, and many went with them over to the Republican Party.[30]

Consistent with the neoconservative ethos of "rollback" during the Cold War, neoconservatives in the 1990s argued against containing Saddam Hussein's regime and instead for its outright removal. Neoconservatives most famously called for the use of force to remove the Hussein regime in an open letter to then-President Bill Clinton. In this letter, dated January 26, 1998, its signatories implored President Clinton "to turn" his "Administration's attention to implementing a strategy for removing Saddam's regime from power." The authors of this letter further held that "the U.S. has the authority under existing UN resolutions to take the necessary steps, *including military steps*, to protect *our vital* interests in the [Persian] Gulf." The letter concludes with the following paragraph:

> *We urge you to act decisively* [on the issue of Iraq]. If you act now to end the threat of weapons of mass destruction against the U.S. or its allies, you will be acting in the most fundamental national security interests of the country. If we accept a course of weakness and drift, we put our interests and our future at risk.[31]

The letter was sent under the auspices of the Project for a New American Century (PNAC), a neoconservative (economic elite–led) policy-planning organization.[32]

Signers of this letter were placed within key foreign policy-making positions within the George W. Bush administration. Prominent among them was Donald Rumsfeld, who as secretary of defense directly oversaw the Iraq invasion. Of the other seventeen individuals who signed the PNAC Iraq letter, ten were given foreign policy–making positions early in the Bush administration: Elliot Abrams (senior director for Near East, Southwest Asian and North African affairs on the National Security Council), James Woolsey (Defense Policy Board), Paula Dobriansky (undersecretary of state for global affairs), Zalmay Khalilzad (president's special envoy to Afghanistan and ambassador at large for free Iraqis), Peter W. Rodman (assistant secretary of defense for international security affairs), William Schneider, Jr.

(chair of the Pentagon's Defense Science Board), Robert B. Zoellick (U.S. trade representative), Richard Armitage (undersecretary of state), Richard Perle (chair of the Defense Policy Board), Paul Wolfowitz (undersecretary of defense), and John Bolton (undersecretary of arms control and international security).[33]

The most significant PNAC member in the Bush administration, especially on the question of Iraq, was Vice President Dick Cheney. Cheney was a founding member of the PNAC along with Rumsfeld.[34] Important for this discussion is the fact that Cheney came to the Bush presidential ticket from Halliburton, where he was president and chief executive officer (since 1995). Halliburton is a multibillion-dollar oil industry firm providing oil field services. Writer and longtime *Los Angeles Times* reporter James Mann reports in his biography (*Rise of the Vulcans*) of leading foreign policy–making officials within the George W. Bush administration that "the invasion of Iraq was in many ways Dick Cheney's war." Mann notes that "within the top ranks [of the U.S. government] the vice president had been the leading proponent of a war to oust Saddam Hussein from power."[35] Mann drew this conclusion after conducting "well over a hundred interviews," many with high-ranking officials within the Bush administration.[36]

Apart from Dick Cheney's economic elite status (as head of Halliburton) and ties (to other oil industry executives), the PNAC's links to economic elites included Donald Rumsfeld, who because of his service in corporate America has accumulated a fortune estimated to be between $50 million and $200 million.[37] The PNAC is also integrated into the conservative policy-planning network through the American Enterprise Institute (AEI)—a conservative think tank.[38] According to Gary Dorrien, a historian of the neoconservative movement:

> The PNAC was closely linked to the American Enterprise Institute, from which it rented office space, and with which it shared a vital connection to the Lynde and Harry Bradley Foundation. Between 1995 and 2001, AEI took in $14.5 million from the Milwaukee-based Bradley Foundation, and PNAC got $1.8 million.[39]

The PNAC also worked jointly with another conservative policy group (think tank), the Heritage Foundation.[40]

Asia-Pacific Partnership for Clean Development and Climate

The Asia-Pacific Partnership for Clean Development and Climate was established in 2005 by the governments of India, China, the United States, Australia, and South Korea. The grouping of these countries in a climate change organization is particularly significant because the United States and China together emit more than 40 percent of total anthropogenic carbon dioxide. Additionally, it is in China and India where most growth in carbon dioxide emissions is expected to occur in the near future.[41] Also, when the partnership was formed, the signatories were outside the emission restrictions of the one international agreement on global warming—the Kyoto Protocol. (Australia has since agreed to be an Annex I country under the protocol and restrict its greenhouse gas emissions accordingly.)[42] Since 2007, the partnership also includes Canada and Japan. According to the partnership's website, "The seven partner countries collectively account for more than half of the world's economy, population and energy use, and they produce about 65 percent of the world's coal, 62 percent of the world's cement, 52 percent of world's aluminum, and more than 60 percent of the world's steel."[43]

The partnership's purpose is to facilitate the sharing of energy technology that could serve to reduce greenhouse gas emissions. As explained by one official at the founding of the organization, "the new pact would aim to use the latest technologies to limit emissions and would try to make sure the technologies were available to countries that need them most."[44] The authors of its website explain that the "partnership will focus on expanding investment and trade in cleaner energy technologies, goods and services in key market sectors. . . . The Partners have approved eight public-private sector task forces covering":

- Aluminum

- Buildings and appliances

- Cement

- Cleaner fossil energy

- Coal mining

- Power generation and transmission

- Renewable energy and distributed generation
- Steel[45]

It is significant that none of these task forces looks at the issue of conservation to address global warming and/or energy depletion.

The partnership's Renewable Energy and Distributed Generation Task Force holds that "renewable energy technologies, such as hydro (large and mini), solar, geothermal, wind and tidal can deliver power with virtually zero emissions." Additionally, "the wide scale deployment of renewable energy and distributed generation technologies increases the diversity of energy supply, and can contribute to improving energy security."[46] This task force is overseeing a number of solar energy research projects: "Building Critical Mass for Ultra High Efficiency Solar Power Stations," "APP Mega Solar Project," "Solar Photovoltaic Linear Concentrator Systems," and "Design and Development of a Solar Biomass Hybrid Cooling and Power Generation System."[47] These projects are geared toward the following:

> The emerging nature of many renewable energy technologies means that there can be market and technical impediments to their uptake, such as cost-competitiveness, awareness of technology options, intermittency and the need for electricity storage. Work is currently being undertaken by many members of the Partnership to address these barriers to increase the wide-scale uptake of renewable energy.[48]

US-China Clean Energy Forum

The *New York Times* describes the US-China Energy Forum (operating since 2008) as "a discussion group of Chinese energy officials and former American cabinet officials."[49] Carol Browner, until recently the Obama administration coordinator for energy and climate change policy,[50] in a message to the forum's president said "that the climate and energy challenge we face cannot be addressed without the committed engagement of our two nations. By being there you are demonstrating the leadership and cooperation that is needed to reduce greenhouse gas emissions, diversify our energy sources, and foster new energy technologies."[51]

The forum's website explains "that China's delegation [to the forum] is led by the National Development and Reform Commission (Han Wenke, Co-Chair, Director General of the Energy Research Institute), and includes prominent leaders from Chinese businesses in energy and finance." The website lists all of the "leaders from the US side" and provides a brief background on them:

Stanley H. Barer is Chairman Emeritus of Saltchuk Resources, a national shipping company, and a Regent of the University of Washington. He is of counsel at Garvey Shubert Barer law firm and was general counsel of the Senate Commerce Committee.

Lawrence Biondi, SJ is president of St. Louis University and a board member of several local and national organizations.

Dennis Bracy is the Chief Executive Officer of the Clean Energy Forum, and Chairman of Avatar Studios, a company that produces television programs and websites for clients around the world, including pioneering television programming in China.

Ambassador William E. Brock is a former Secretary of Labor, U.S. Trade Representative and Senator from Tennessee.

Ambassador Carla Hills is chairman and chief executive officer of Hills & Company and formerly was US Trade Representative and Secretary of Housing and Urban Development. She serves as co-chairman of the Council on Foreign Relations and chairman of the National Committee on U.S.-China Relations.

Ambassador Mickey Kantor is a partner in the law firm of Mayer, Brown, Rowe & Maw. He previously served as U.S. Secretary of Commerce and as U.S. Trade Representative.

Secretary Norman Mineta is vice chairman of Hill & Knowlton and is a former Secretary of Commerce and

Transportation, Member of Congress and Mayor of San Jose, California.

Sharon Nelson is former chair of Consumers Union, board member of the National Commission on Energy Policy and the North American Electric Reliability Corporation.

Ambassador J. Stapleton Roy is Director, Kissinger Institute on China and the United States, at the Woodrow Wilson International Center for Scholars and was US Ambassador to China.

Susan Tierney is a member of the Obama energy transition team, on leave as managing principal of the Analysis Group and chair of the Energy Foundation. She formerly was Assistant Secretary of the Department of Energy.

Ambassador Darryl Johnson has had a distinguished career in the U.S. State Department. He is a former U.S. Ambassador to Lithuania, Thailand, Philippines, and a former director of the American Institute of Taiwan.

The Forum's website goes on to add that "the Forum is supported by a wide variety of leading businesses and NGOs [nongovernment agencies], each a leader in its sector":

Sponsors include GE, Russell Investment Group, the US-China Business Council, AMCHAM China, AMCHAM Shanghai, Stark Investments, Garvey Schubert Barer, Itron, Hill & Knowlton, Puget Sound Energy, McKinsey & Company, Babcock & Wilcox, Better Place, University of Washington, Battelle, The Energy Foundation, Pacific Gas & Electric, Starbucks, Applied Materials, Washington State University, Consolidated Edison Company of New York, The Boeing Company, Avatar Studios, the Port of Seattle, Stellar International Networks, Sapphire Energy and the Woodrow Wilson International Center for Scholars, State Grid, China Southern Grid.[52]

The forum's self-described "mission is to recommend specific and transformational actions to the governments of the United States and China to improve energy efficiency, accelerate the deployment of clean energy and thereby reduce carbon emissions." In seeking to accomplish these goals, the forum in 2009 set out eight initiatives, including "Establish a Joint [U.S.-China] Energy Laboratory" (which recently has been agreed to[53]); "Increase efficiency, development and commercialization of Solar Photovoltaic, Solar Thermal, and Concentrated Solar technologies. Set ambitious goals for solar deployment by 2020"; and "Power the transformation to Electric Vehicles."[54]

The WBCSD

The WBCSD is made up of a number of global corporations that are headquartered all over the world. In 1996, the WBCSD had about 125 corporate members drawn from eight regions of the globe: Western Europe, Central/Eastern Europe, Africa/Middle East, North America, Latin America, Japan, Asia, and Oceania. Among its members are Renault, Total, Volkswagen, Fiat, Statoil (Norway/Petroleum), Volvo, British Petroleum, Shell Oil, Texaco, Mitsubishi, and Toyota.[55] According to its website, the WBCSD is a "CEO-led, global association of some 200 companies dealing exclusively with business and sustainable development." Its "members are drawn from more than 30 countries and 20 major industrial sectors." The WBCSD also "benefits from a global network of some 60 national and regional business councils and regional partners."[56]

The WBCSD posits itself as a proponent of the ecological modernization of global capitalism. At the core of ecological modernization is the idea that environmental protection and economic growth are complementary goals. This complementary relationship can be achieved through the development and application of technology and environmentally safer products.[57]

In its mission statement, the WBCSD casts its primary purpose in terms consistent with ecological modernization theory. The council's mission "is to provide business leadership as a catalyst for change toward sustainable development, and to support the business license to operate, innovate and grow in a world increasingly shaped by sustainable development issues."[58] Additionally, the WBCSD views itself as "a platform for companies to explore sustainable development, share knowledge, experiences and best practices, and to advocate business positions on these issues in a variety of forums, working with governments, non-governmental and intergovernmental organizations."[59] Moreover, in a 1996 publication, the WBCSD declares that "society must take a longer view and create the necessary frameworks to reward business for adding ever more value while using fewer resources and producing less pollution."[60]

The WBCSD, however, advocates a narrow or "weak" conception of ecological modernization. A narrow approach to ecological modernization relies heavily on technological solutions and alternative fuels to address natural resource depletion.[61] In a 2005 document titled *Pathways to 2050: Energy and Climate Change*, the WBCSD targets specific advice for different regions around the world on energy. In the case of the United States, no mention is made of its automobile dependency, nor does the WBCSD counsel explicit conservation programs. Instead, the WBCSD advises the United States that by 2050 there will be "A 100% improvement in vehicle efficiency, large scale use of biomass fuels and the growth of hydrogen fuel cell vehicles to more than one-quarter of the on road fleet." Additionally, there will be "a restart in nuclear power growth yielding a 40% increase in capacity" and "large-scale use of renewables, especially wind and solar."[62]

A more expansive or "strong" conception of ecological modernization would involve ecologically sensitive land management. This type of land management would entail the intensive usage of land (as opposed to sprawl), drawing residential and work areas closer together, and creating smaller work and living spaces in urban areas. Ecologically sensitive land management would move residents away from their dependence on the automobile (and the internal combustion engine) and toward more energy-conserving forms of transportation, such as walking, bicycling, and mass transit.[63]

Hence, the objective of the WBCSD with regard to the United States is not necessarily to reduce the energy throughputs in the operation of its economy to minimize energy use, but to put urban sprawl in America on an economically and environmentally sustainable path—specifically through the development and deployment of more efficient technologies, nuclear power, and alternative fuels. (The former tactic is indicative of strong ecological modernization.) The shortcomings in the WBCSD's approach to energy depletion (at least in the United States) are twofold. First, technological solutions could simply serve to shift the economic stress (i.e., resource depletion) created by one activity from one aspect of the economy onto another.[64] The case of nuclear power demonstrates this point. Although greater reliance on nuclear power will reduce fossil fuel use, the production of nuclear power and the disposal of nuclear wastes both have significant environmental liabilities, which in the long term could be more economically damaging than energy depletion.[65] Perhaps most menacing, increased U.S. reliance on nuclear power also could increase the possibility of nuclear war, as this could encourage other countries to pursue/expand civilian nuclear power programs. Such programs can be utilized for nuclear weapons production.[66] Another economically hazardous substitute for fossil fuels is ethanol. It is a liquid fuel derived from organic material (e.g., corn) that can be used to power automobiles, trucks, and buses.[67] As corn, soy, and other food products have been diverted to ethanol production, the price of food has significantly increased.[68]

The second difficulty associated with a technological approach to the reducing supply of fossil fuels (especially oil), which includes an approach rooted in alternative fuels, is that no technology has been forthcoming to effectively confront the question of fuel fossil depletion (or global warming) within the current context of global capitalism.[69] In other words, no technology to date has been developed to allow current rates of economic growth and consumption to continue without the externalities of energy depletion and global warming (or, alternatively, the creation of intractable amounts of nuclear waste).

Hydrogen, for instance, has been posited (by the WBCSD among others[70]) as a potentially clean, unlimited, and affordable replacement for fossil fuels. A scientist of the Natural Resources Defense Council, however, opined about hydrogen fuel use that "real revolutions have to occur before this is going to become a large-scale reality." He went on to note, "It very possibly could happen, but" a hydrogen-based economy is "not a sure thing."[71] Writing in the journal *Nature*, Brian C. H. Steele, of the University of London's Imperial College, and Angelika Heinzel, of the University of Duisburg-Essen in Germany, concluded that "unless there is a breakthrough in the production of hydrogen and the development of new hydrogen-storage materials, the concept of a 'hydrogen economy' will remain an unlikely scenario."[72] While we wait for such breakthroughs, the economic and environmental effects of global warming and energy shortfalls could irreversibly come to a head.

The WBCSD and Fixed-Rail Transport

As noted earlier, the WBCSD in a 2005 document does not advise that the United States (even by 2050) adopt fixed-rail transport on a large scale in its urban zones. Nevertheless, on the issue of "Mobility," the WBCSD does predict "by 2050" a "shift to mass transportation," which "offers considerable efficiency benefits." The WBCSD counsels the use of biomass, hydrogen, hybrids, and diesel throughout the years until 2045. In "2050," however, the WBCSD foresees that "substantial investments are made worldwide to make [fixed-rail transportation] an efficient and attractive alternative to individual transport."[73]

Similarly, in a 2004 WBCSD report titled *Mobility 2030: Meeting the Challenge of Sustainability* (authored by executives from General Motors, Toyota, Shell Oil, British Petroleum, DaimlerChrysler, Ford, Honda, Michelin Tire Company, Nissan Motor, Norsk Hydro, Renault, and Volkswagen), the following observation is offered:

> Over the very long run—*five decades or more*—societies face a fundamental choice about how their mobility patterns will develop. Some hold that in order to make mobility sustainable, people will have to be induced to live in significantly more dense agglomerations. According to this view, only by doing this will it be technologically and financially feasible to rely on public transport to a much greater degree than is generally the case today.[74]

In opposition to those who champion efficient land use and greatly expanded public mass transit, the authors of *Mobility 2030* declare:

> To us, this strategy seems to rest on forcing people to adapt to the technological and economic characteristics of transport systems. An alternative strategy is to adapt the technological and economic characteristics of transport systems to fit the living choices of the public. The various

vehicle technologies we have described [throughout the
Mobility 2030 report] appear to have the potential to enable
such an adaptation.[75]

The ICC

In contrast to the WBCSD, the ICC has a broad membership base.
According an ICC brochure, it "groups hundreds of thousands of
member companies and associations from over 130 countries."[76] The
ICC's website goes on to explain that it

> is the voice of world business championing the global
> economy as a force for economic growth, job creation
> and prosperity. . . . ICC activities cover a broad spectrum,
> from arbitration and dispute resolution to making the case
> for open trade and the market economy system, business
> self-regulation, fighting corruption or combating commer-
> cial crime.[77]

Among ICC members are AT&T, Chevron, Citigroup, DuPont,
ExxonMobil, Fiat, Ford, General Electric, General Motors, J.P. Mor-
gan Chase, Nissan Motor, Norsk Hydro, Procter & Gamble, Sony,
and Toyota.[78]

The ICC conducts much of its policy work through commissions.
Its website notes that its "commissions are the bedrock of ICC." They
are "composed of a total of more than 500 business experts who give
freely of their time to formulate ICC policy and elaborate its rules."
The author(s) of this website go on to explain that ICC "commissions
scrutinize proposed international and national government initiatives
affecting their subject areas and prepare business positions for submis-
sion to international organizations and governments."[79]

One of the ICC's sixteen commissions is the Commission on
Environment and Energy. It "is comprised of 227 members represent-
ing 75 multinational corporations as well as representatives from 33
industry associations, and 52 ICC national committees that federate
ICC members in their countries." The Commission on Environment
and Energy "examines major environmental and energy related policy
issues of interest to world business via task forces and thematic groups.

The Commission usually meets twice a year though task forces and other thematic groups may meet more frequently." The author(s) of the ICC website hold that "Commission members gain influence at the national level through the ICC's global network of national committees and at the international level through ICC's privileged links with major intergovernmental organisations."[80]

The ICC commission on energy notes on its website that "access to reliable, affordable, economically viable, socially acceptable and environmentally sound energy is fundamental to economic growth and sustainable development." More specifically, the "ICC supports energy technology development," as "a broad variety of energy resources and technologies will be required to meet the varying needs of individual countries or markets." Moreover, "energy security is a vital consideration not only for day-to-day operations, but also for long-term investment." In addition, the ICC energy commission holds that "energy efficiency is another critical component of any comprehensive sustainable energy strategy."[81] In a 2009 report on energy efficiency, one of the commission's "key messages" was that "increas[ing] energy efficiency can make a significant impact in squaring the circle between an increased demand for energy and . . . ensuring a move towards a more sustainable energy future."[82]

On its website, the energy commission explains that the "ICC participated in the United Nations Commission for Sustainable Development (UNCSD) process through the Business Action for Energy (BAE), a temporary business network that facilitated business input to UNCSD." Another participant in the BAE is the WBCSD.[83]

Conclusion

Ever since U.S. petroleum consumption made it onto the political agenda (with the 1973 oil shock), economic elite policy groups have advocated mostly supply side answers to liquid energy shortages. The Twentieth Century Fund international oil and energy task forces in the 1970s argued that the U.S. government should work to bring more petroleum into the world oil market. Vice President Cheney's National Energy Policy Development Group (in consultation with major energy firms) posited the same in the early 2000s. The Cheney-led Project for a New American Century in the late 1990s held that the

United States should invade Iraq, the country with the second largest proven oil reserves. (At the time, Cheney was an oil executive.) Under Cheney's leadership in 2003, the United States did exactly that.

A secondary position put forward by the Twentieth Century Fund task forces of the 1970s was for the United States to respond to the volatility in the global petroleum market with efficiency initiatives and research into alternative fuels. The U.S. government tended to ignore this advice, with minimal public investments in alternative energy (including solar), the winding down of nuclear power, and the American SUV craze of the 1980s and beyond.[84] In the contemporary era, the Asia-Pacific Partnership for Clean Development and Climate, the US-China Clean Energy Forum, the WBCSD, and the ICC have publicly embraced technological efficiency and alternative energy (including nuclear).

The United States remains highly dependent on fossil fuels, as well as exposed and vulnerable to the global petroleum market. This exposure and vulnerability flows predominately from urban sprawl. Government and economic elite policy discussion groups generally avoid discussing the oil dependency resulting from low-density urban development (i.e., urban sprawl). Although the WBCSD does indicate that a worldwide expansion in fixed-rail urban transit will ultimately occur, it does not foresee such a shift beginning until 2045 and does not predict that it will occur in the United States (the country with ostensibly the most automobile-dependent and oil-profligate urban zones in the world[85]). Moreover, the WBCSD produced a report in 2004 in which a number of its members expressed hostility to the notion of a global movement mandating high-density urban development and energy-conserving mass transportation.

Conclusion

Urban Sprawl, Energy, and the Failure of Empire

One of the ironies—some would say tragedies—of the U.S. alternative energy program of the 1950s, 1960s, and 1970s is that the government's own scientists advised against an ambitious civilian nuclear energy program. The AEC's General Advisory Committee (made up of leading nuclear scientists) in the late 1940s held that there were too many uncertainties surrounding nuclear power to pursue it at a large scale. More than sixty years after the General Advisory Committee first reported against civilian nuclear power, these uncertainties remain. First, nuclear power is extremely dangerous to produce. Although nuclear power safety measures have evolved, the chances of a catastrophic accident continue to persist, especially as plants age.[1] (Japan's Fukushima Daiichi multiple nuclear meltdowns struck a political blow to nuclear power worldwide[2]—with Germany publicly announcing the phasing out of its nuclear power capacity by 2022 and Japan's prime minister calling for an end to nuclear power in Japan.[3]) Another unresolved nuclear power–related risk is nuclear waste. There is no way to safely store nuclear waste for the entirety of its radioactive life. Additionally, nuclear waste can be converted into weapons-grade material. Finally, the expense of building nuclear plants makes nuclear power an economically questionable source of energy.

One argument in this book is that the oil shocks of the 1970s brought the intent (and politics) underlying U.S. energy policy into sharp relief. This is so in two ways. One, in the face of the oil shocks, the United States somewhat counterintuitively stopped ordering new nuclear plants. This serves to negate the idea that the United States pursued nuclear energy as a way to sure up its domestic supply of energy. The initial intent actually underlying nuclear energy in the United States was to solidify its hegemonic global position. By the late 1970s, it was evident that nuclear power would not serve this purpose, and the United States began to wind down its nuclear program. Conversely, European countries sought to expand their nuclear power capacity as a means to enhance national autonomy on the issue of energy in the face of OPEC's pricing power and the selective Saudi Arabia oil embargo. France was particularly successful in this regard (Chapter 5).

Two, in spite of the oil shocks, the United States maintains the most oil-dependent urban zones in the world.[4] Instead of curbing urban sprawl and its automobile dependency, the United States in the aftermath of the 1970s oil shocks focused its foreign policy on the Persian Gulf region, the area with half of the world's proven petroleum reserves (Chapter 5). In 2003, the United States went as far as invading and occupying Iraq, the country with reputedly the second largest proven reserves of oil (Chapter 6). This seemingly confirms the political and economic priority afforded to urban sprawl by the leaders of the American polity.

Of course, nuclear power is not a direct replacement for petroleum (i.e., gasoline), but it could have been. Nuclear power plants could be linked to a robust urban fixed-rail transit system. This is what Europe (especially France) has done. Additionally, research could have been financed to study the possibility of converting nuclear-derived electricity into liquid fuel (i.e., hydrogen) or electrically powered vehicles. Similarly, save for a short stint in the late 1970s (during the Iranian Revolution), the United States historically did not make solar energy research a political priority. Even today, in the era of global warming, the government of China makes much greater efforts (and investments) to develop and deploy solar energy infrastructure than the United States.[5]

In the end, the United States (particularly its urban zones) is highly dependent on the world oil market—consuming roughly 20 million barrels of petroleum per day. This poses two specific threats. One is the issue of energy depletion. The International Energy Agency, a body that advises almost all of the advanced governments of the world, recently concluded that global conventional petroleum production peaked in 2006.[6] This means that oil supply worldwide will slowly decline. With demand for oil expected to remain stable or increase (China's automobile fleet, for instance, is growing substantially[7]), this slow decline will likely translate into ever climbing petroleum prices. Two, the Middle East is where the majority of the world's oil is located. The political instability of this region (e.g., the recent "Arab Spring"[8]) greatly adds to the energy insecurity and financial vulnerability derived from the United States' extreme oil automobile/dependency. (The United States increased this instability with its invasion and occupation of Iraq.)

U.S. oil/automobile dependency has broad global implications. First, residents of the United States consume about 20 percent of the world's goods and services. A dramatic spike in oil prices, or a major disruption in petroleum supply, will send the world economy into depression/recession.[9] Of even more profound concern is global warming. While the United States insists on consuming large amounts of fossil fuels through its sprawled urban zones, no international strategy/agreement on climate change is seemingly possible. Global warming science indicates that humanity does not have much time left before anthropogenic carbon dioxide emissions lead to catastrophic climate change.[10]

With the United States facing the dilemmas of global warming, oil depletion, and global political instability, the American government has taken initial steps to build additional nuclear power plants and made tentative moves toward solar energy. Whether these efforts to more greatly diversify the American economy away from fossil fuels are in time (or sufficient) to prevent the worst aspects of petroleum exhaustion and/or global warming from striking the United States and the world, only time will tell.[11] What is definite is that U.S. historic urban sprawl and energy policies have steered the world toward a climate and economic precipice.

Consistent with economic elite theory (Chapter 2), I have argued throughout this book that these policies have been driven by economic elites through policy discussion groups: the Rockefeller Foundation (Chapter 3), the Panel on the Impact of the Peaceful Uses of Atomic Energy (Chapter 3), the Association for Applied Solar Energy (Chapter 3), the President's Conference on Unemployment (of the early 1920s) (Chapter 4), the President's Emergency Committee on Housing (of

1934) (Chapter 4), the Twentieth Century Fund task forces of the
1970s (one on energy and the other on the international oil crisis)
(Chapter 5), Vice President Cheney's National Energy Policy Development Group (Chapter 6), the Project for a New American Century
(of the late 1990s) (Chapter 6), the WBCSD (Chapter 6), and the
ICC (Chapter 6). Collectively, these groups, and the economic elites
who have led them, historically championed U.S. urban sprawl (and
continue to do so) as well as nuclear energy as a foreign policy device
and opposed major government support for solar energy.

Notes

Chapter 1

1. George A. Gonzalez, *Urban Sprawl, Global Warming, and the Empire of Capital* (Albany: State University of New York Press, 2009).

2. John Houghton, *Global Warming: The Complete Briefing*, 4th ed. (New York: Cambridge University Press, 2009); Mark Maslin, *Global Warming: A Very Short Introduction* (New York: Oxford University Press, 2009).

3. Harvey Blatt, *America's Environmental Report Card: Are We Making the Grade?*, 2nd ed. (Cambridge, MA: MIT Press, 2011), 181.

4. For the purpose of brevity, when using the term solar power, I am referring to passive solar, photovoltaic, wind, and wave power.

5. Political scientist Frank N. Laird notes that "government interest in solar technologies during most of the 1950s and 1960s showed up mainly in the military services and the space program, both of which had small research programs that were mostly concerned with photovoltaics (PVs), devices that convert light directly into electricity." *Solar Energy, Technology Policy, and Institutional Values* (New York: Cambridge University Press, 2001), 52. Also see John Perlin, *From Space to Earth: The Story of Solar Electricity* (Cambridge, MA: Harvard University Press, 2002), chap. 5.

6. In the warmer and desert regions of the United States, passive (or thermal) solar hot water heaters in homes have historically been used. More specifically, in the early twentieth century, sales of home solar water heaters surged in southern California, but by 1920 this market all but disappeared because of the availability of inexpensive natural gas and the incentives natural gas companies gave to purchasers of gas-powered home water heaters. In the 1930s, there was a burgeoning market for home solar water heaters in Florida, but it was effectively undermined in the postwar period by cheap electricity and the high labor and material costs of solar water heaters. Ken Butti and John Perlin, *A Golden Thread: 2500 Years of Solar Architecture and Technology* (Palo Alto, CA: Cheshire Books, 1980), chaps. 11–12; Alexis

Madrigal, *Powering the Dream: The History and Promise of Green Technology* (Cambridge, MA: Da Capo Press, 2011), chap. 11.

7. Robert Righter, *Windfall: Wind Energy in America Today* (Norman: University of Oklahoma Press, 2011); Matthew L. Wald, "Taming Unruly Wind Power," *New York Times*, Nov. 5, 2011, B1.

8. Felicity Barringer, "With Push toward Renewable Energy, California Sets Pace for Solar Power," *New York Times*, July 16, 2009, A19; Todd Woody, "Solar Power Projects Face Potential Hurdles," *New York Times*, Oct. 29, 2010, B1; Sara Hamdan, "Energy Plant Makes a Leap in Solar Power," *New York Times*, Oct. 25, 2011. Web.

9. Laird, *Solar Energy, Technology Policy, and Institutional Values*; Joseph J. Romm, *The Hype about Hydrogen: Fact and Fiction in the Race to Save the Climate* (Washington, DC: Island Press, 2004); Travis Bradford, *Solar Revolution: The Economic Transformation of the Global Energy Industry* (Cambridge, MA: MIT Press, 2006); Bruce Podobnik, *Global Energy Shifts: Fostering Sustainability in a Turbulent Age* (Philadelphia: Temple University, Press, 2006); Ion Bogdan Vasi, *Winds of Change: The Environmental Movement and the Global Development of the Wind Energy Industry* (New York: Oxford University Press, 2011); Heather Timmons and Vikas Bajaj, "Emerging Economies Move Ahead with Nuclear Plans," *New York Times*, Mar. 15, 2011, B1; Kate Galbraith, "Wind Power Gains as Gear Improves," *New York Times*, Aug. 7, 2011. Web; Josh Wingrove, "Can Coal Come Clean or Is Wind the Future?" *Globe and Mail*, Sept. 13, 2011, A3; Michelle Michot Foss, "Ignorance Stifles Innovation in Solving Energy Problems," *International Herald Tribune*, Oct. 12, 2011, Finance sec., 11; Matthew Wald, "Batteries at a Wind Farm Help Control Output," *New York Times*, Oct. 29, 2011, B3; Paul Krugman, "Here Comes the Sun," *New York Times*, Nov. 7, 2011, A25; Keith Bradsher, "China Bends to U.S. Complaint on Solar Panels but Plans Retaliation," *New York Times*, Nov. 22, 2011, B7; Vikas Bajaj, "India's Investment in the Sun," *New York Times*, Dec. 29, 2011, B1; Richard K. Lester and David M. Hart, *Unlocking Energy Innovation* (Cambridge, MA: MIT Press, 2012); Matthew L. Wald, "Storehouses for Solar Energy Can Step In When the Sun Goes Down," *New York Times*, Jan. 3, 2012, B1; James Kanter, "Obstacles to Danish Wind Power," *International Herald Tribune*, Jan. 23, 2012, Finance sec., 18.

10. David Elliot, ed., *Nuclear or Not? Does Nuclear Power Have a Place in a Sustainable Energy Future?* (New York: Palgrave MacMillan, 2007); John M. Broder, "U.S. Nuclear Industry Faces New Uncertainty," *New York Times*, Mar. 14, 2011, A1; Aubrey Belford, "Indonesia to Continue Plans for Nuclear Power," *New York Times*, Mar. 18, 2011, B5; Charles D. Ferguson, *Nuclear Energy: What Everyone Needs to Know* (New York: Oxford University Press, 2011); Catherine Hornby, "Italy Plans to Reassess Nuclear Power in Few Years," Reuters, Apr. 26, 2011.

China has made the most aggressive recent move into civilian nuclear power. Nonetheless, by 2020, even with its planned spurt of nuclear reactor construction, China would generate less than 10 percent of its electricity from nuclear. Keith Bradsher, "Nuclear Power Expansion in China Stirs Concerns," *New York Times*, Dec. 16, 2009, A1; Keith Bradsher, "A Radical Kind of Reactor," *New York Times*, Mar. 25, 2011, B1; Matthew L. Wald, "N.R.C. Lowers Estimate of How Many Would Die in Meltdown," *New York Times*, July 30, 2011, A14.

11. Matthew L. Wald, "New Interest in Turning Gas to Diesel," *New York Times*, Dec. 24, 2010, B1.

12. Andrew Jacobs, "Chinese and British Officials Tangle in Testy Exchange over Climate Agreement," *New York Times*, Dec. 23, 2009, A10; John M. Broder, "Climate Talks End with Modest Deal on Emissions," *New York Times*, Dec. 12, 2010, A16.

13. Matthew L. Wald, "Study of Baby Teeth Sees Radiation Effects," *New York Times*, Dec. 14, 2010, D2; Kirk Johnson, "A Battle over Uranium Bodes Ill for U.S. Debate," *New York Times*, Dec. 27, 2010, A1.

14. David E. Sanger and Matthew L. Wald, "Radioactive Releases in Japan Could Last Months, Experts Say," *New York Times*, Mar. 14, 2011, A1; Hiroko Tabuchi and Matthew L. Wald, "Partial Meltdowns Presumed at Crippled Reactors," *New York Times*, Mar. 14, 2011, A7; Ken Belson, Keith Bradsher, and Matthew L. Wald, "Executives May Have Lost Valuable Time at Damaged Nuclear Plant," *New York Times*, Mar. 20, 2011, A12; Hiroko Tabuchi, "Japan Passes Law Supporting Stricken Nuclear Plant's Operator," *New York Times*, Aug. 4, 2011, A8.

15. Matthew L. Wald, "Administration to Push for Small 'Modular' Reactors," *New York Times*, Feb. 13, 2011, A29.

16. Duncan Lyall Burn, *Nuclear Power and the Energy Crisis: Politics and the Atomic Industry* (New York: New York University Press, 1978); Gabrielle Hecht, *The Radiance of France: Nuclear Power and National Identity after World War II* (Cambridge, MA: MIT Press, 1998 [2009]); Matthew Wald, "Sluggish Economy Curtails Prospects for Building Nuclear Reactors," *New York Times*, Oct. 11, 2010, B1; Bob Herbert, "A Price Too High?," *New York Times*, Mar. 19, 2011, A23; Ian Austen, "Uranium Processor Still Optimistic about Nuclear Industry," *New York Times*, Mar. 26, 2011, B3.

17. The federal government was seeking a fee of $880 million on a loan guarantee of $7.6 billion. Matthew L. Wald, "Fee Dispute Hinders Plan for Reactor," *New York Times*, Oct. 10, 2010, A21.

18. Ellen Barry, "Lessons From Chernobyl for Japan," *New York Times*, Mar. 20, 2011, WK1; Andrew E. Kramer, "Nuclear Industry in Russia Sells Safety, Taught by Chernobyl," *New York Times*, Mar. 23, 2011, B1.

19. Deborah Guber, *The Grassroots of a Green Revolution: Polling America on the Environment* (Cambridge, MA: MIT Press, 2003).

20. Michael Cooper and Dalia Sussman, "Nuclear Power Loses Support in New Poll," *New York Times*, Mar. 23, 2011, A15; Hiroko Tabuchi and Keith Bradsher, "Japan Nuclear Disaster Put on Par with Chernobyl," *New York Times*, Apr. 12, 2011, A10.

21. John R. Logan and Harvey L. Molotch, *Urban Fortunes: The Political Economy of Place* (Berkeley: University of California Press, 1987 [2007]); George A. Gonzalez, *The Politics of Air Pollution: Urban Growth, Ecological Modernization, and Symbolic Inclusion* (Albany: State University of New York Press, 2005).

22. Rick Eckstein, *Nuclear Power and Social Power* (Philadelphia: Temple University Press, 1997).

23. Dorothy Nelkin and Michael Pollak, *The Atom Besieged: Extraparliamentary Dissent in France and Germany* (Cambridge, MA: MIT Press, 1981); Jim Falk, *Global Fission: The Battle over Nuclear Power* (New York: Oxford University Press, 1982); Alain Touraine, *Anti-nuclear Protest: The Opposition to Nuclear Energy in France* (New York: Cambridge University Press, 1983); Brian Balogh, *Chain Reaction: Expert Debate and Public Participation in American Commercial Nuclear Power, 1945–1975* (New York: Cambridge University Press, 1991); John Wills, *Conservation Fallout: Nuclear Protest at Diablo Canyon* (Reno: University of Nevada Press, 2006).

24. Matthew L. Wald, "Vermont Nuclear Plant Up for Sale," *New York Times*, Nov. 5, 2010, B9; and "Showdown on Vermont Nuclear Plant's Fate," *New York Times*, Mar. 11, 2011, A19.

25. "N.J. Nuclear Plant Closing Early," Associated Press, Dec. 8, 2010.

26. Luther J. Carter, *Nuclear Imperatives and Public Trust: Dealing with Radioactive Waste* (Washington, DC: Resources for the Future, 1987); Richard Burleson Stewart and Jane Bloom Stewart, *Fuel Cycle to Nowhere: U.S. Law and Policy on Nuclear Waste* (Nashville, TN: Vanderbilt University Press, 2011); Matthew L. Wald, "A Safer Nuclear Crypt," *New York Times*, July 6, 2011, B1.

27. Robert Vandenbosch and Susanne E. Vandenbosch, *Nuclear Waste Stalemate: Political and Scientific Controversies* (Salt Lake City: University of Utah Press, 2007); Max S. Power, *America's Nuclear Wastelands: Politics, Accountability, and Cleanup* (Pullman: Washington State University Press, 2008); J. Samuel Walker, *The Road to Yucca Mountain: The Development of Radioactive Waste Policy in the United States* (Berkeley: University of California Press, 2009); Matthew L. Wald, "Report Urges Storing Spent Nuclear Fuel, Not Reprocessing It," *New York Times*, Apr. 26, 2011, A16; John M. Broder and Matthew L. Wald, "Report Blasts Management Style of Nuclear Regulatory Commission Chairman," *New York Times*, June 11, 2011, A13; Matthew L.

Wald, "Court Won't Intervene in Fate of Nuclear Dump," *New York Times*, July 2, 2011, A13; Kate Galbraith, "A New Urgency to the Problem of Storing Nuclear Waste," *International Herald Tribune*, Nov. 28, 2011, Finance sec., 20.

28. Irvin C. Bupp and Jean-Claude Derian, *The Failed Promise of Nuclear Power: The Story of Light Water* (New York: Basic Books, 1978).

29. Matthew Kroenig, *Exporting the Bomb: Technology Transfer and the Spread of Nuclear Weapons* (Ithaca, NY: Cornell University Press, 2010); Jo Becker and William J. Broad, "New Doubts about Turning Plutonium into a Fuel," *New York Times*, Apr. 11, 2011, A14.

30. Richard G. Hewlett and Jack M. Holl, *Atoms for Peace and War 1953–1961: Eisenhower and the Atomic Energy Commission* (Berkeley: University of California Press, 1989); Ira Chernus, *Eisenhower's Atoms for Peace* (College Station: Texas A&M University Press, 2002).

31. Michael J. Brenner, *Nuclear Power and Non-Proliferation: The Remaking of U.S. Policy* (New York: Cambridge University Press, 1981); Mark Hertsgaard, *Nuclear Inc.: The Men and Money behind Nuclear Energy* (New York: Pantheon, 1983); Joseph A. Camilleri, *The State and Nuclear Power: Conflict and Control in the Western World* (Seattle: University of Washington Press, 1984), 252–257; Robert L. Beckman, *Nuclear Non-Proliferation, Congress, and the Control of Peaceful Nuclear Activities* (Boulder, CO: Westview Press, 1985); J. Michael Martinez, "The Carter Administration and the Evolution of American Nuclear Nonproliferation Policy, 1977–1981," *Journal of Policy History* 14, no. 3 (2002): 261–292.

32. Rajiv Nayan, *The Nuclear Non-Proliferation Treaty and India* (New York: Routledge, 2011); Vikas Bajaj, "Resistance to Jaitapur Nuclear Plant Grows in India," *New York Times*, Apr. 14, 2011, B1.

33. Mark Mazzetti, "U.S. Intelligence Finding Says Iran Halted Its Nuclear Arms Effort in 2003," *New York Times*, Dec. 4, 2007, A1.

34. Ian Bellany, Coit D. Blacker, and Joseph Gallacher, *The Nuclear Non-proliferation Treaty* (New York: Routledge, 1985); Sverre Lodgaard, *Nuclear Disarmament and Non-Proliferation: Towards a Nuclear-Weapon-Free World?* (New York: Routledge, 2010); Olav Njølstad, *Nuclear Proliferation and International Order: Challenges to the Non-Proliferation Treaty* (New York: Routledge, 2010).

35. Dana Allin and Steven Simon, *The Sixth Crisis: Iran, Israel, America, and the Rumors of War* (New York: Oxford University Press, 2010); David E. Sanger, "U.S. and Allies Plan More Sanctions against Iran," *New York Times*, Dec. 11, 2010, A6; Jo Becker, "U.S. Approved Business with Blacklisted Nations," *New York Times*, Dec. 24, 2010, A1; David E. Sanger, "Iran Moves to Shelter Its Nuclear Fuel Program," *New York Times*, Sept. 2, 2011, A4.

36. Judy Dempsey, "Panel Urges Germany to Close Nuclear Plants by 2021," *New York Times*, May 12, 2011, B7. The United States is currently

building four nuclear reactors and completing one that had remained unfinished. Matthew L. Wald, "Alabama Nuclear Reactor, Partly Built, to Be Finished," *New York Times*, Aug. 19, 2011, A12.

37. Norman Polmar, *U.S. Nuclear Arsenal: A History of Weapons and Delivery Systems since 1945* (Annapolis, MD: Naval Institute Press, 2009); Jerry Miller, *Stockpile: The Story behind 10,000 Strategic Nuclear Weapons* (Annapolis, MD: Naval Institute Press, 2010); Peter Baker, "Arms Talks Now Turn to Short-Range Weapons," *New York Times*, Dec. 25, 2010, A4.

38. William Yong, "Gas Prices Soar in Iran as Subsidy Is Reduced," *New York Times*, Dec. 20, 2010, A6.

39. Kenneth S. Deffeyes, *Hubbert's Peak: The Impending World Oil Shortage* (Princeton, NJ: Princeton University Press, 2001); *Beyond Oil: The View from Hubbert's Peak* (New York: Hill and Wang, 2005); and *When Oil Peaked* (New York: Hill and Wang, 2010); David Goodstein, *Out of Gas: The End of the Age of Oil* (New York: Norton, 2004); Paul Roberts, *The End of Oil: On the Edge of a Perilous New World* (New York: Houghton Mifflin, 2004); Richard Heinberg, *The Party's Over: Oil, War, and the Fate of Industrial Societies*, 2nd ed. (Gabriola Island, BC: New Society Publishers, 2005); Jeremy K. Leggett, *Half Gone: Oil, Gas, Hot Air, and the Global Energy Crisis* (London: Portobello, 2005); John R. Moroney, *Power Struggle: World Energy in the Twenty-First Century* (Westport, CT: Praeger, 2008); Stephen Kurczy, "International Energy Agency Says 'Peak Oil' Has Hit," *Christian Science Monitor*, Nov. 11, 2010; Paul Krugman, "The Finite World," *New York Times*, Dec. 27, 2010, A19; Michael J. Graetz, *The End of Energy: The Unmaking of America's Environment, Security, and Independence* (Cambridge, MA: MIT Press, 2011).

40. Steve Isser, *The Economics and Politics of the United States Oil Industry, 1920–1990: Profits, Populism, and Petroleum* (New York: Routledge, 1996); Roger M. Olien and Diana Davids Olien, *Oil and Ideology: The Cultural Creation of the American Petroleum Industry* (Chapel Hill: University of North Carolina Press, 2000); Diana Davids Olien and Roger M. Olien, *Oil in Texas: The Gusher Age, 1895–1945* (Austin: University of Texas Press, 2002); Francisco Parra, *Oil Politics: A Modern History of Petroleum* (New York: I. B. Tauris, 2004).

41. Richard H. Vietor, *Environmental Politics and the Coal Coalition* (College Station: Texas A&M University Press, 1980); and *Energy Policy in America since 1945* (New York: Cambridge University Press, 1984); Barbara Freese, *Coal: A Human History*, 4th ed. (New York: Penguin, 2004); Marc Humphries, ed., *U.S. Coal: A Primer on the Major Issues* (Hauppauge, NY: Novinka Books, 2004); Jeff Goodell, *Big Coal: The Dirty Secret behind America's Energy Future* (New York: Mariner, 2007); Christopher F. Jones,

"A Landscape of Energy Abundance: Anthracite Coal Canals and the Roots of American Fossil Fuel Dependence, 1820–1860," *Environmental History* 15, no. 3 (2010): 449–484; 54; Blatt, *America's Environmental Report Card*, 2nd ed., 158.

42. M. Elizabeth Sanders, *The Regulation of Natural Gas: Policy and Politics, 1938–1978* (Philadelphia: Temple University Press, 1981); Paul W. MacAvoy, *The Natural Gas Market: Sixty Years of Regulation and Deregulation* (New Haven, CT: Yale University Press, 2001); David G. Victor, Amy M. Jaffe, and Mark H. Hayes, eds., *Natural Gas and Geopolitics: From 1970 to 2040* (New York: Cambridge University Press, 2006); Kate Galbraith, "Natural Gas, Scrutinized, Pushes for Growth," *New York Times*, Mar. 11, 2011, A21; Jad Mouawad, "Natural Gas Now Viewed as Safer Bet," *New York Times*, Mar. 22, 2011, B1; "The Future of Natural Gas," *The Economist*, Aug. 6, 2011; Ian Urbina, "New Report by Agency Lowers Estimates of Natural Gas in U.S.," *New York Times*, Jan. 29, 2012, A16.

43. John G. Ikenberry, *Reasons of State: Oil Politics and the Capacities of American Government* (Ithaca: Cornell University Press, 1988); Simon Bromley, *American Hegemony and World Oil: The Industry, the State System and the World Economy* (University Park: Pennsylvania State University Press, 1991); Daniel Yergin, *The Prize: The Epic Quest for Oil, Money, and Power* (New York: Simon & Schuster, 1991); David Davis, *Energy Politics* (New York: St. Martin's, 1993); Rachel Bronson, *Thicker Than Oil: America's Uneasy Partnership with Saudi Arabia* (New York: Oxford University Press, 2006); Bruce A. Beaubouef, *The Strategic Petroleum Reserve: U.S. Energy Security and Oil Politics, 1975–2005* (College Station: Texas A&M University Press, 2007); John S. Duffield, *Over a Barrel: The Costs of U.S. Foreign Oil Dependence* (Stanford: Stanford University Press, 2008); Daniel Yergin, *The Quest: Energy, Security, and the Remaking of the Modern World* (New York: Penguin, 2011).

44. Laird, *Solar Energy, Technology Policy, and Institutional Values*.

45. Emmanuel Todd, *After the Empire: The Breakdown of the American Order*, trans. C. Jon Delogu (New York: Columbia University Press, 2003); Michael H. Hunt, *The American Ascendancy: How the United States Gained and Wielded Global Dominance* (Chapel Hill: University of North Carolina Press, 2007).

46. Peter S. Goodman, "The Economy: Trying to Guess What Happens Next," *New York Times*, Nov. 25, 2007, sec. 4, p. 1.

47. Gonzalez, *Urban Sprawl, Global Warming, and the Empire of Capital*; and "An Eco-Marxist Analysis of Oil Depletion via Urban Sprawl," *Environmental Politics* 15, no. 4 (2006): 515-531; Steven Erlanger, "With Prospect of U.S. Slowdown, Europe Fears a Worsening Debt Crisis," *New York Times*, Aug. 8, 2011, B3.

48. Michael Kamber and Taimoor Shag, "Iran Stops Fuel Delivery, Afghanistan Says, and Prices Are Rising," *New York Times*, Dec. 23, 2010, A12.

49. Michael T. Hatch, *Politics and Nuclear Power: Energy Policy in Western Europe* (Lexington: University Press of Kentucky, 1986); Torleif Haugland, Helge Ole Bergensen, and Kjell Roland, *Energy Structures and Environmental Futures* (New York: Oxford University Press, 1998); Simon Romero, "Oil-Rich Norwegians Take World's Highest Gasoline Prices in Stride," *New York Times*, Apr. 30, 2005, C1; Gonzalez, *Urban Sprawl, Global Warming, and the Empire of Capital*, chap. 5.

50. Ellen Meiksins Wood, *The Origin of Capitalism* (New York: Monthly Review Press, 1999); and *Empire of Capital* (New York: Verso, 2003).

51. Daniel H. Nexon and Thomas Wright, "What's at Stake in the American Empire Debate," *American Political Science Review* 101, no. 2 (2007): 253–271.

52. William S. Borden, *The Pacific Alliance: United States Foreign Economic Policy and Japanese Trade Recovery, 1947–1955* (Madison: University of Wisconsin Press, 1984); Michael Schaller, *Altered States: The United States and Japan since the Occupation* (New York: Oxford University Press, 1997); Gary Herrigel, *Industrial Constructions: The Sources of German Industrial Power* (New York: Cambridge University Press, 2000); Horst Siebert, *The German Economy: Beyond the Social Market* (Princeton, NJ: Princeton University Press, 2005); John Swenson-Wright, *Unequal Allies? United States Security and Alliance Policy toward Japan, 1945–1960* (Stanford, CA: Stanford University Press, 2005); Belay Seyoum, *Export-Import Theory, Practices, and Procedures*, 2nd ed. (New York: Routledge, 2008).

53. Philip J. Funigiello, *American-Soviet Trade in the Cold War* (Chapel Hill: University of North Carolina Press, 1988); Lewis H. Siegelbaum, *Cars for Comrades: The Life of the Soviet Automobile* (Ithaca, NY: Cornell University Press, 2008).

54. Francis Shor, *Dying Empire: U.S. Imperialism and Global Resistance* (New York: Routledge, 2010).

55. Clifford Krauss, "As Ethanol Takes Its First Steps, Congress Proposes a Giant Leap," *New York Times*, Dec. 18, 2007, C1; John M. Broder, "House, 314–100, Passes Broad Energy Bill," *New York Times*, Dec. 19, 2007, A24; Matthew L. Wald, "U.S. Backs Project to Produce Fuel from Corn Waste," *New York Times*, July 7, 2011, B10; Clifford Krauss, "Ethanol Subsidies Besieged," *New York Times*, July 8, 2011, B1; Clifford Krauss, "Here's an Easy $100 Billion Cut: Ending the Tax Subsidies for Oil and Ethanol Is Fiscally Sound and Right," *New York Times*, Aug. 8, 2011, A18; Matthew L. Wald, "A Fine for Not Using a Biofuel That Doesn't Exist," *New York Times*, Jan. 10, 2012, B1.

56. Harvey Blatt, *America's Environmental Report Card*, 2nd ed., 180–181; Robert B. Semple, Jr., "Oil and Gas Had Help. Why Not Renewables?" *New York Times*, Oct. 16, 2011, SR10.

57. Joseph A. Camilleri, *The State and Nuclear Power*.

58. John Krige, *American Hegemony and the Postwar Reconstruction of Science in Europe* (Cambridge, MA: MIT Press, 2006).

59. Duncan Burn, *Nuclear Power and the Energy Crisis* (New York: New York University Press, 1978); Jim Falk, *Global Fission: The Battle over Nuclear Power* (New York: Oxford University Press, 1982); John L. Campbell, *Nuclear Power and the Contradiction of U.S. Policy* (Ithaca, NY: Cornell University Press, 1988); James M. Jasper, *Nuclear Politics: Energy and the State in the United States, Sweden, and France* (Princeton, NJ: Princeton University Press, 1990); Terrence Price, *Political Electricity: What Future for Nuclear Energy?* (New York: Oxford University Press, 1990); Helen Caldicott, *Nuclear Power Is Not the Answer* (New York: New Press, 2006).

60. In 1954, the Atomic Energy Commission reported to the Congressional Joint Committee on Atomic Energy that

> the United States could offer to export either heavy-water or light-water reactors under the Atoms-for-Peace program. Heavy-water reactors might be more attractive to European nations because they could probably obtain supplies of heavy water and natural uranium without depending on the United States. If, however, the United States selected light-water reactors for export, the [Atomic Energy] Commission would have to supply the slightly enriched uranium fuel because no European countries were likely to make the heavy financial commitment necessary to build an enrichment plant. One advantage, then, of using light-water reactors for export . . . was that the United States could control both the supply of uranium fuel elements and also reprocessing of spent fuel. Hewlett and Holl, *Atoms for Peace and War*, 197–198.

61. Camilleri, *The State and Nuclear Power*, 193.

62. Brenner, *Nuclear Power and Non-Proliferation*; Hertsgaard, *Nuclear Inc.*, 75–79; Camilleri describes the 1973 change in U.S. enriched nuclear fuel policy in the following:

> Whereas previously customers could obtain through "requirement contracts" as much enriched fuel as they needed with less than one year's advance notice, they were now expected to sign "fixed commitment" contracts at least eight years in advance of delivery of the first core. In addition, they had to agree to purchase

specified amounts of enriched uranium over a moving ten-year period and to deliver the required uranium supplies regardless of their actual need for enriched fuel. . . . The net effect of the new contract system was to shift the risks to the user. *The State and Nuclear Power*, 195.

63. Paul L. Joskow, "The International Nuclear Industry Today: The End of the American Monopoly," *Foreign Affairs* 54, no. 4 (1976): 788–803; Brenner, *Nuclear Power and Non-Proliferation*; Camilleri, *The State and Nuclear Power*, 193–202.

64. Nuclear fuel processing involves uranium enrichment and the recycling of nuclear waste to produce/extract more nuclear fuel from said waste.

65. Brenner, *Nuclear Power and Non-Proliferation*; Hertsgaard, *Nuclear Inc.*, chap. 9; Beckman, *Nuclear Non-Proliferation*; Martinez, "The Carter Administration and the Evolution of American Nuclear Nonproliferation Policy."

66. William J. Broad, "For Iran, Enriching Only Gets Easier," *New York Times*, Mar. 8, 2010, D1; David E. Sanger and William J. Broad, "Iran Says It Will Speed Up Uranium Enrichment," *New York Times*, June 9, 2011, A14; David E. Sanger and William J. Broad, "Survivor of Attack Leads Nuclear Effort in Iran," *New York Times*, July 23, 2011, A4.

67. Camilleri notes that "by the late 1970s, the proliferation of enrichment techniques was already well advanced." *The State and Nuclear Power*, 197; also see David Albright, Frans Berkhout, and William Walker, *Plutonium and Highly Enriched Uranium, 1996: World Inventories, Capabilities, and Policies* (New York: Oxford University Press, 1997); Matthew L. Wald, "Loan Request by Uranium-Enrichment Firm Upends Politics as Usual," *New York Times*, Nov. 25, 2011, B5.

68. Political scientist Michael T. Hatch writes in *Politics and Nuclear Power* that "past German dependence on American enriched uranium has given the United States great potential leverage," 133. Camilleri writes in more general terms when he explains that through the provision of nuclear fuel, the United States sought to "impose a highly visible form of extraterritorial control over [its allies'] economic and foreign policies. If nothing else it called into question the independence of their national [nuclear] programs." *The State and Nuclear Power*, 256.

Concerns over sovereignty and independence of national nuclear programs were piqued with the American Nuclear Non-Proliferation Act of 1978. With this legislation, the Carter administration attempted to use the U.S. provision of nuclear fuel and technology to determine European (and other countries') domestic and foreign nuclear policies. Brenner, *Nuclear Power and Non-Proliferation*; Camilleri, *The State and Nuclear Power*,

255–257; Hatch, *Politics and Nuclear Power*, 123–134; Martinez, "The Carter Administration and the Evolution of American Nuclear Nonproliferation Policy," 272–273.

69. William J. Broad, "Buffett Helps Create Nuclear Fuel Bank," *New York Times*, Dec. 4, 2010, A4; William J. Broad and David E. Sanger, "C.I.A. Secrets Could Surface in Swiss Nuclear Case," *New York Times*, Dec. 24, 2010, A1; John Mueller, *Atomic Obsession* (New York: Oxford University Press, 2010); Catherine Collins and Douglas Frantz, *Fallout: The True Story of the CIA's Secret War on Nuclear Trafficking* (New York: Free Press, 2011).

70. Jesse S. Tatum, *Energy Possibilities: Rethinking Alternatives and the Choice-Making Process* (Albany, State University of New York Press, 1995).

71. Richard Pfau, *No Sacrifice Too Great: The Life of Lewis L. Strauss* (Charlottesville: University Press of Virginia, 1984), 193; Hewlett and Holl, *Atoms for Peace and War*, 194–195.

72. Peter Newman, Timothy Beatley, and Heather Boyer, *Resilient Cities: Responding to Peak Oil and Climate Change* (Washington, DC: Island Press, 2009); Elisabeth Rosenthal, "African Huts Far from the Grid Glow with Renewable Power," *New York Times*, Dec. 25, 2010, A1; James Kanter, "A Solar and Wind Revolution from a Land of Oil," *New York Times*, Mar. 13, 2011. Web; Allison Gregor, "Idle Land Finds a Purpose as Farms for Solar Power," *New York Times*, Mar. 23, 2011, B7.

73. Keith Bradsher, "China Racing Ahead of U.S. in the Drive to Go Solar," *New York Times*, Aug. 25, 2009, A1; Todd Woody, "G.E. Plans to Build Largest Solar Panel Plant in U.S.," *New York Times*, Apr. 7, 2011, B3; Keith Bradsher, "China Charges Protectionism in Call for Solar Panel Tariffs," *New York Times*, Oct. 22, 2011, B6.

74. Andrew E. Kramer, "Safety Issues Linger as Nuclear Reactors Shrink in Size," *New York Times*, Mar. 19, 2010, B1.

75. Andrew E. Kramer, "Russia Is Seeking to Build Europe's Nuclear Plants," *New York Times*, Oct. 12, 2010, B11; and "Nuclear Industry in Russia Sells Safety"; Trevor Findlay, *Nuclear Energy and Global Governance* (New York: Routledge, 2011), 12.

76. Hertsgaard, *Nuclear Inc.*, 282–283.

77. Keith Bradsher, "China Leading Global Race to Make Clean Energy," *New York Times*, Jan. 30, 2010, A1; Keith Bradsher, "On Clean Energy, China Skirts Rules," *New York Times*, Sept. 9, 2010, A1; Clifford Krauss, "In Global Forecast, China Looms Large as Energy User and Maker of Green Power," *New York Times*, Nov. 10, 2010, B3; Sewell Chan, "U.S. Says China Fund Breaks Rules," *New York Times*, Dec. 23, 2010, B1; Keith Bradsher, "Solar Panel Maker Moves Work to China," *New York Times*, Jan. 15, 2011, B1; Keith Bradsher, "U.S. Posted a Trade Surplus in Solar Technologies, Study Finds," *New York Times*, Aug. 29, 2011, B4; Keith Bradsher,

"China Benefits as U.S. Solar Industry Withers," *New York Times*, Sept. 2, 2011, B1; Keith Bradsher, "200 Chinese Subsidies Violate Rules, U.S. Says," *New York Times*, Oct. 7, 2011, B3; Keith Bradsher, "U.S. Solar Firms Accuse Chinese of Trade Violations," *New York Times*, Oct. 20, 2011, B1.

Harvey Blatt reports that "China is already the world's leading renewable energy producer. . . . Its investment in renewable energy increased from $352 million in 2007 to $736 million in 2008 to $34.6 billion in 2009." China "has announced plans to spend $75 billion a year on clean energy." Blatt, *America's Environmental Report Card*, 2nd ed., 184.

78. Keith Schneider, "Midwest Emerges as Center for Clean Energy," *New York Times*, Dec. 1, 2010, B8.

79. Matthew L. Wald, "U.S. Backs New Loans for Projects on Energy," *New York Times*, Sept. 29, 2011, A14; Kate Galbraith, "Future of Solar and Wind Power May Hinge on Federal Aid," *New York Times*, Oct. 26, 2011, F5; Eric Lipton and Clifford Krauss, "A Gold Rush of Subsidies in the Search for Clean Energy," *New York Times*, Nov. 12, 2011, A1.

80. Eric Lipton and Clifford Krauss, "A U.S.-Backed Geothermal Plant in Nevada Struggles," *New York Times*, Oct. 3, 2011, B1; Jackie Calmes, "Leader Picked for Review of U.S. Loans on Energy," *New York Times*, Oct. 29, 2011, A15. In 2009, the U.S. Congress authorized the use of $70 billion in grants, loans, and loan guarantees "to award for high-tech research and commercial projects for renewable energy." Also in 2009, government funding for "solar energy research" was $175 million. Blatt, *America's Environmental Report Card*, 2nd ed., 180–181.

According to the Environmental Law Institute, from 2002 to 2008 the U.S. federal government allocated $12.2 billion in subsidies for "traditional renewables" (wind, solar, hydrogen, geothermal, biomass [excluding corn ethanol], hydro). During this period of time, $16.8 billion in federal subsidies went to corn ethanol. Graph: Energy Subsidies Black, Not Green. Also see Environmental Law Institute, Estimating U.S. Government Subsidies to Energy Sources: 2002-2008 (Environmental Law Institute, 2009). This graph and study can be found at: www.eli.org/pressdetail.cfm?ID=205.

81. Matt Daily and Sarah McBride, "Financing Dearth Holds Solar Back in U.S.," *New York Times*, Oct. 17, 2010. Web.

82. Tim Büthe, "Taking Temporality Seriously: Modeling History and the Use of Narratives as Evidence," *American Political Science Review* 96, no. 3 (2002): 481–493.

83. In 2002, Germany enacted a law mandating the shutdown of all its nuclear power plants by 2022. Judy Dempsey, "Germany Extends Nuclear Plants' Life," *New York Times*, Sept. 6, 2010. Web.

84. Dempsey, "Germany Extends Nuclear Plants' Life."

In response to Japan's Fukushima nuclear disaster in March 2011, the German government shut down the country's oldest eight nuclear power plants and reverted back to the original deadline of phasing out Germany's nuclear energy capacity by 2022. Elisabeth Rosenthal, "Germany Dims Nuclear Plants, but Hopes to Keep Lights On," *New York Times*, Aug. 30, 2011, A1.

85. Matthew L. Wald, "U.S. Rejects Nuclear Plant over Design of Key Piece," *New York Times*, Oct. 16, 2009, A13.

86. Kate Galbraith, "Certainties of 1970s Energy Crisis Have Fallen Away," *New York Times*, Apr. 3, 2011, A25.

Chapter 2

1. Robert A. Dahl and Charles E. Lindblom, *Politics, Economics, and Welfare* (New Haven, CT: Yale University Press, 1953); Robert A. Dahl, *A Preface to Democratic Theory* (Chicago: University of Chicago Press, 1956); Robert A. Dahl, *Who Governs? Democracy and Power in an American City* (New Haven, CT: Yale University Press, 1961 [2005]); Sheldon Kamieniecki, *Corporate America and Environmental Policy: How Often Does Business Get Its Way?* (Palo Alto, CA: Stanford University Press, 2006); Frank R. Baumgartner, Jeffrey M. Berry, Marie Hojnacki, David C. Kimball, and Beth L. Leech, *Lobbying and Policy Change: Who Wins, Who Loses, and Why* (Chicago: University of Chicago Press, 2009); Thomas T. Holyoke, *Competitive Interests: Competition and Compromise in American Interest Group Politics* (Washington, D.C.: Georgetown University Press, 2011).

2. David B. Truman, *The Governmental Process: Political Interests and Public Opinion* (New York: Knopf, 1951).

3. Theodore J. Lowi, *The End of Liberalism: The Second Republic of the United States* (New York: Norton, 1979).

4. Grant McConnell, *Private Power and American Democracy* (New York: Knopf, 1966).

5. Robert A. Dahl and Charles E. Lindblom, preface to *Politics, Economics, and Welfare* (New Haven, CT: Yale University Press, 1976), xxxvii.

6. Arthur Selwyn Miller, *The Modern Corporate State: Private Governments and the American Constitution* (Westport, CT: Greenwood, 1976).

7. The Federal Housing Authority is the unofficial name of the Federal Housing Administration.

8. Marc Weiss, *The Rise of the Community Builders: The American Real Estate Industry and Urban Land Planning* (New York: Columbia University Press, 1987); George A. Gonzalez, *Urban Sprawl, Global Warming, and the Empire of Capital* (Albany: State University of New York, 2009), chap. 4.

9. Weiss, *The Rise of the Community Builders*; Gonzalez, *Urban Sprawl, Global Warming, and the Empire of Capital*.

10. Joan Hoff Wilson, *American Business & Foreign Policy, 1920-1933* (Lexington: University Press of Kentucky, 1971), chap. 7.

11. E.g., Robert J. Duffy, *Nuclear Politics in America: A History and Theory of Government Regulation* (Lawrence: University of Kansas Press, 1997); Franj N. Von Hippel, "It Could Happen Here," *New York Times*, Mar. 24, 2011, A31; Eric Lichtblau, "Lobbyists' Long Effort to Revive Nuclear Industry Faces New Test," *New York Times*, Mar. 25, 2011, A1. Otherwise, numerous books and articles on the politics of U.S. civilian nuclear power are cited in Chapter 1. Virtually all of this literature describes the special interest politics of the nuclear power industry.

12. Mark Hertsgaard, *Nuclear Inc.: The Men and Money behind Nuclear Energy* (New York: Pantheon, 1983).

13. Hertsgaard, *Nuclear Inc.*, 177.

14. Frank G. Dawson, *Nuclear Power: Development and Management of a Technology* (Seattle: University of Washington Press, 1976).

15. The United States ended the recycling of nuclear waste/fuel for use in civilian power plants in 1977. Joseph A. Camilleri, *The State and Nuclear Power* (Seattle: University of Washington Press, 1984), 212.

16. Robert Vandenbosch and Susanne E. Vandenbosch, *Nuclear Waste Stalemate: Political and Scientific Controversies* (Salt Lake City: University of Utah Press, 2007); Max S. Power, *America's Nuclear Wastelands: Politics, Accountability, and Cleanup* (Pullman: Washington State University Press, 2008); J. Samuel Walker, *The Road to Yucca Mountain: The Development of Radioactive Waste Policy in the United States* (Berkeley: University of California Press, 2009).

17. Rick Eckstein, *Nuclear Power and Social Power* (Philadelphia: Temple University Press, 1997).

18. John R. Logan and Harvey L. Molotch, *Urban Fortunes: The Political Economy of Place* (Berkeley: University of California Press, 1987 [2007]); George A. Gonzalez, *The Politics of Air Pollution: Urban Growth, Ecological Modernization, and Symbolic Inclusion* (Albany: State University of New York Press, 2005).

19. David A. Kirsch, *The Electric Vehicle and the Burden of History* (New Brunswick, NJ: Rutgers University Press, 2000); Joseph J. Romm, *The Hype about Hydrogen: Fact and Fiction in the Race to Save the Climate* (Washington, DC: Island Press, 2004); William J. Mitchell, Christopher E. Borroni-Bird, and Lawrence D. Burns, *Reinventing the Automobile: Personal Urban Mobility for the 21st Century* (Cambridge, MA: MIT Press, 2010).

20. Panel on the Impact of the Peaceful Uses of Atomic Energy, *Peaceful Uses of Atomic Energy*, vol. 1 (Washington, DC: Government Printing Office, 1956), xi.

21. Panel on the Impact of the Peaceful Uses of Atomic Energy, *Peaceful Uses of Atomic Energy*, vol. 2 (Washington, DC: Government Printing Office, 1956), xiii–xviii.

22. Panel on the Impact of the Peaceful Uses of Atomic Energy, *Peaceful Uses of Atomic Energy*, vol. 1, p. 2.

23. I offer a full description this model elsewhere (George A. Gonzalez, *Corporate Power and the Environment: The Political Economy of U.S. Environmental Policy* (Lanham, MD: Rowman & Littlefield, 2001), 10–13), so here I provide only its central features.

24. Theda Skocpol, *States and Social Revolutions* (Cambridge: Cambridge University Press, 1979); Eric A. Nordlinger, *On the Autonomy of the Democratic State* (Cambridge, MA: Harvard University Press, 1981); Stephen Skowronek, *Building a New American State: The Expansion of National Administrative Capacities, 1877–1920* (Cambridge: Cambridge University Press, 1982); Theda Skocpol, "Bringing the State Back In: Strategies of Analysis in Current Research," in Peter Evans, Dietrich Rueschemeyer, and Theda Skocpol, eds., *Bringing the State Back In* (Cambridge: Cambridge University Press, 1985); Daniel P. Carpenter, *The Forging of Bureaucratic Autonomy: Reputations, Networks, and Policy Innovations in Executive Agencies, 1862–1928* (Princeton, NJ: Princeton University Press, 2001).

25. Adam Rome, *The Bulldozer in the Countryside: Suburban Sprawl and the Rise of American Environmentalism* (Cambridge: Cambridge University Press, 2001), chap. 1; also see Gail Radford, *Modern Housing for America: Policy Struggles in the New Deal Era* (Chicago: University of Chicago Press, 1996).

26. Stephen Krasner, *Defending the National Interest: Raw Materials Investments and U.S. Foreign Policy* (Princeton, NJ: Princeton University Press, 1978); also see John G. Ikenberry, *Reasons of State: Oil Politics and the Capacities of American Government* (Ithaca, NY: Cornell University Press, 1988).

27. Theda Skocpol, *Protecting Soldiers and Mothers: The Political Origins of Social Policy in the United States* (Cambridge, MA: Harvard University Press, 1992); Theda Skocpol, Marshall Ganz, and Ziad Munson, "A Nation of Organizers: The Institutional Origins of Civic Voluntarism in the United States," *American Political Science Review* 94, no. 3 (2000): 527–546.

28. Patrick McGrath, *Scientists, Business, and the State, 1890–1960* (Chapel Hill: University of North Carolina Press, 2002); Andrew Rich, *Think Tanks, Public Policy, and the Politics of Expertise* (New York: Cambridge University Press, 2004); Judith A. Layzer, "Deep Freeze: How Business Has Shaped the Global Warming Debate in Congress," in *Business and Environmental Policy*, Michael E. Kraft and Sheldon Kamieniecki, eds. (Cambridge, MA: MIT Press, 2007).

29. Theda Skocpol, "A Brief Response [to G. William Domhoff]," *Politics and Society* 15, no. 3 (1986/87): 332.

30. Seth Shulman, *Undermining Science: Suppression and Distortion in the Bush Administration* (Berkeley: University of California Press, 2006); John Heilprin, "White House Rejects Mandatory CO2 Caps," Associated Press,

Feb. 2, 2007; Catherine Gautier, *Oil, Water, and Climate: An Introduction* (New York: Cambridge University Press, 2008).

31. Catherine Gautier and Jean-Louis Fellous, eds., *Facing Climate Change Together* (New York: Cambridge University Press, 2008); John T. Houghton, *Global Warming: The Complete Briefing*, 4th ed. (New York: Cambridge University Press, 2009); James Lawrence Powell, *The Inquisition of Climate Science* (New York: Columbia University Press, 2011).

32. Donald A. Brown, *American Heat: Ethical Problems with the United States' Response to Global Warming* (Lanham, MD: Rowman and Littlefield, 2002); Michael Lisowski, "Playing the Two-Level Game: US President Bush's Decision to Repudiate the Kyoto Protocol," *Environmental Politics* 11, no. 4 (2002): 101–119; Dana Fisher, *National Governance and the Global Climate Change Regime* (Lanham, MD: Rowman and Littlefield, 2004), chap. 6; Loren R. Cass, *The Failures of American and European Climate Policy: International Norms, Domestic Politics, and Unachievable Commitments* (Albany: State University of New York Press, 2006); William Nordhaus, *A Question of Balance* (Cambridge, MA: MIT Press, 2008).

33. Andrew Jordan, Dave Huitema, Harro van Asselt, Tim Rayner, and Frans Berkhout, eds., *Climate Change Policy in the European Union: Confronting the Dilemmas of Mitigation and Adaptation?* (New York: Cambridge University Press, 2010).

34. John M. Broder and Elisabeth Rosenthal, "Obama Has Goal to Wrest a Deal in Climate Talks," *New York Times*, Dec. 18, 2009, A1; John M. Broder, "Director of Policy on Climate Will Leave, Her Goal Unmet," *New York Times*, Jan. 25, 2011, A15; John M. Broder, "House Panel Votes to Strip E.P.A. of Power to Regulate Greenhouse Gases," *New York Times*, Mar. 11, 2011, A18. In 2011, the *New York Times* reported: "Many countries fault the United States for government inaction on climate change." Cornelia Dean, "Group Urges Research into Aggressive Efforts to Fight Climate Change," *New York Times*, Oct. 4, 2011, A18.

35. Harvey Strum, "Eisenhower's Solar Energy Policy," *Public Historian* 6, no. 2 (1984): 37–50.

36. *Proceedings of the World Symposium on Applied Solar Energy, Phoenix, AZ, Nov. 1–5, 1955* (San Francisco: Jorgenson & Co., 1956); Farrington Daniels, *Direct Use of the Sun's Energy* (New Haven, CT: Yale University Press, 1964); Harvey Strum, "Eisenhower's Solar Energy Policy."

37. Frank N. Laird, *Solar Energy, Technology Policy, and Institutional Values* (New York: Cambridge University Press, 2001), 53.

38. Harvey Strum and Fred Strum, "American Solar Energy Policy, 1952–1982," *Environmental Review* 7 (summer 1983): 147. Also see William D. Metz and Allen L. Hammond, *Solar Energy in America* (Washington, DC: American Association for the Advancement of Science, 1978) and Laird, *Solar Energy*, 166–167.

39. Strum and Strum, "American Solar Energy Policy, 1952–1982," 136.

40. Strum and Strum, "American Solar Energy Policy, 1952–1982," 141.

41. Strum and Strum, "American Solar Energy Policy, 1952–1982." Also see Alexis Madrigal, *Powering the Dream: The History and Promise of Green Technology* (Cambridge, MA: DaCapo Press, 2011), chap. 13.

42. Strum and Strum, "American Solar Energy Policy, 1952–1982," 150.

43. John F. Manley, "Neo-pluralism: A Class Analysis of Pluralism I and Pluralism II," *American Political Science Review* 77 no. 2 (1983): 368–383.

44. Ralph Miliband, *The State in Capitalist Society* (New York: Basic Books, 1969); Colin Hay, Michael Lister, and David Marsh, eds., *The State: Theories and Issues* (New York: Palgrave Macmillan, 2006), chap. 2; Fred Block, "Understanding the Diverging Trajectories of the United States and Western Europe: A Neo-Polanyian Analysis," *Politics & Society* 35, no. 1 (2007): 3–33; Paul Wetherly, Clyde W. Barrow, and Peter Burnham, eds., *Class, Power and the State in Capitalist Society: Essays on Ralph Miliband* (New York: Palgrave MacMillan, 2008).

45. Clyde W. Barrow, *Critical Theories of the State* (Madison: University of Wisconsin Press, 1993), 17; Barrow explains that "corporations emerged as the dominant economic institutions in capitalist societies by the end of the nineteenth century." He goes on to note that as early as the late 1920s, "the bulk of U.S. economic activity, whether measured in terms of assets, profits, employment, investment, market shares, or research and development, was concentrated in the fifty largest financial institutions and five hundred largest nonfinancial corporations." Ibid.

Political scientists Jeffrey A. Winters and Benjamin I. Page, writing in 2009, hold that "it is now appropriate to . . . think about the possibility of *extreme* political inequality, involving great political influence by a very small number of extremely wealthy individuals." They add that "we argue that it is useful to think about the U.S. political system in terms of oligarchy." Jeffrey A. Winters and Benjamin I. Page, "Oligarchy in the United States," *Perspectives on Politics* 7, no. 4 (2009): 744, emphasis in original. Also see Paul Krugman, "Oligarchy, American Style," *New York Times*, Nov. 4, 2011, A31, and Shaila Dewan and Robert Gebeloff, "One Percent, Many Variations," *New York Times*, Jan. 15, 2012, A1.

46. Barrow, *Critical Theories of the State*, 16; Mike McIntire, "Nonprofit Acts as a Stealth Business Lobbyist," *New York Times*, April 22, 2012, A1.

47. William Appleman Williams, *The Roots of the Modern American Empire* (New York: Random House, 1969), 98.

48. Ibid.

49. G. William Domhoff, *The Bohemian Grove and Other Retreats* (New York: Harper and Row, 1974).

50. Michael Useem, *The Inner Circle: Large Corporations and the Rise of Business Political Activity in the U.S. and U.K.* (Oxford: Oxford University

Press, 1984); also see Beth Mintz and Michael Schwartz, *The Power Structure of American Business* (Chicago: University of Chicago Press, 1985).

51. G. William Domhoff, *Who Rules America?*, 6th ed. (New York: McGraw-Hill, 2009), chap. 4.

52. G. William Domhoff, *The Powers That Be* (New York: Random House, 1978), 61.

53. James Weinstein, *The Corporate Ideal in the Liberal State: 1900–1918* (Boston: Beacon Press, 1968); David Eakins, "Business Planners and America's Postwar Expansion," in David Horowitz, ed., *Corporations and the Cold War* (New York: Monthly Review Press, 1969); David Eakins, "Policy-Planning for the Establishment," in Ronald Radosh and Murray N. Rothbard, eds., *A New History of Leviathan* (New York: E.P. Dutton & Co., 1972); Gabriel Kolko, *The Triumph of Conservatism: A Reinterpretation of American History, 1900–1916* (New York: Free Press, 1977 [1963]); Domhoff, *The Powers That Be*; Domhoff, *Who Rules America?*, chap. 4.

54. Fred Bunyan Joyner, *David Ames Wells: Champion of Free Trade* (Cedar Rapids, IA: Torch Press, 1939); Walter Lafeber, *The New Empire: An Interpretation of American Expansion, 1860–1898* (Ithaca, NY: Cornell University Press, 1963); Williams, *Roots of the Modern American Empire*; Richard W. Turk, *The Ambiguous Relationship: Theodore Roosevelt and Alfred Thayer Mahan* (New York: Greenwood, 1987); Warren Zimmermann, *First Great Triumph: How Five Americans Made Their Country a World Power* (New York: Farrar, Straus and Giroux, 2002).

55. Weinstein, *Corporate Ideal in the Liberal State*; Eakins, "Business Planners and America's Postwar Expansion"; Domhoff, *The Powers That Be*; G. William Domhoff, *The Power Elite and the State* (New York: Aldine de Gruyter, 1990); Edward H. Berman, *The Ideology of Philanthropy: The Influence of the Carnegie, Ford, and Rockefeller Foundations on American Foreign Policy* (Albany: State University of New York Press, 1983); Clyde W. Barrow, *Universities and the Capitalist State: Corporate Liberalism and the Reconstruction of American Higher Education, 1894–1928* (Madison: University of Wisconsin Press, 1990); Clyde W. Barrow, "Corporate Liberalism, Finance Hegemony, and Central State Intervention in the Reconstruction of American Higher Education," *Studies in American Political Development* 6 (fall 1992): 420–444; Barrow, *Critical Theories of the State*, chap. 1; G. William Domhoff, *State Autonomy or Class Dominance?* (New York: Aldine de Gruyter, 1996); Mark Dowie, *American Foundations: An Investigative History* (Cambridge, MA: MIT Press, 2001); Christopher J. Cyphers, *The National Civic Federation and the Making of New Liberalism, 1900–1915* (Westport, CT: Praeger, 2002); Inderjeet Parmar, "American Foundations and the Development of International Knowledge Networks," *Global Networks* 2, no. 1 (2002): 13–30; Joan Roelofs, *Foundations and Public Policy: The Mask of Pluralism* (Albany:

State University of New York Press, 2003); Domhoff, *Who Rules America?*; G. William Domhoff and Michael J. Webber, *Class and Power in the New Deal* (Palo Alto, CA: Stanford University Press, 2011).

56. Domhoff, *The Powers That Be*, 63.

57. Eakins, "Policy-Planning for the Establishment"; Domhoff, *The Powers That Be*, 61–87; Barrow, *Critical Theories of the State*, chap. 1; Domhoff, *Who Rules America?*, chap. 4; Gonzalez, *Urban Sprawl, Global Warming, and the Empire of Capital*, chap. 6. The economic elite–led policy-planning network has two groupings—one that is characterized as "moderate" or "corporate liberal" and the other as "conservative." Although these two groups will frequently compromise on issues, they sometimes cannot. When they cannot find common ground, their struggles will usually spill over into government, where each will utilize its political strength to try and get its way. Weinstein, *Corporate Ideal in the Liberal State*; Eakins, "Business Planners and America's Postwar Expansion"; Eakins, "Policy-Planning for the Establishment"; Domhoff, *The Powers That Be*, chap. 3; Domhoff, *The Power Elite and the State*, 38–39; Barrow, *Critical Theories of the State*, chap. 1.

58. Mark Dowie, *Losing Ground: American Environmentalism at the Close of the Twentieth Century* (Cambridge, MA: MIT Press, 1995), 58–59; Dowie, *American Foundations*; Roelofs, *Foundations and Public Policy*.

59. Susan R. Schrepfer, *The Fight to Save the Redwoods: A History of Environmental Reform, 1917–1978* (Madison: University of Wisconsin Press, 1983), 10; also see Holway R. Jones, *John Muir and the Sierra Club: The Battle for Yosemite* (San Francisco: Sierra Club, 1965); and Richard J. Orsi, "'Wilderness Saint' and 'Robber Baron': The Anomalous Partnership of John Muir and the Southern Pacific Company for Preservation of Yosemite National Park," *Pacific Historian* 29 (summer–fall 1985): 136–152.

60. Schrepfer, *Fight to Save the Redwoods*, 171–173; also see Michael P. Cohen, *The History of the Sierra Club, 1892–1970* (San Francisco: Sierra Club Books, 1988).

61. Schrepfer, *Fight to Save the Redwoods*, 113; also see George A. Gonzalez, "The Wilderness Act of 1964 and the Wilderness Preservation Policy Network," *Capitalism Nature Socialism* 20, no. 4 (2009): 31–52.

62. Gonzalez, *Corporate Power and the Environment*; George A. Gonzalez, "Ideas and State Capacity, or Business Dominance? A Historical Analysis of Grazing on the Public Grasslands," *Studies in American Political Development* 15 (fall 2001): 234–244; *Politics of Air Pollution*; "The Comprehensive Everglades Restoration Plan: Economic or Environmental Sustainability?," *Polity* 37, no. 4 (2005): 466–490; and "The Wilderness Act of 1964 and the Wilderness Preservation Policy Network."

63. Felicity Barringer, "A Coalition for Firm Limit on Emissions," *New York Times*, Jan. 19, 2007, C1.

64. Samuel Hays, "The Politics of Reform in Municipal Government in the Progressive Era," *Pacific Northwest Quarterly* 55, no. 4 (1964): 157–169; G. William Domhoff, *Who Really Rules: New Haven and Community Power Reexamined* (Santa Monica, CA: Goodyear, 1978).

65. John R. Logan and Harvey L. Molotch, *Urban Fortunes: The Political Economy of Place* (Berkeley: University of California Press, 1987 [2007]), 152.

66. Nico Poulantzas, *Political Power and Social Classes* (London: New Left Books, 1973); Barrow, *Critical Theories of the State*, chap. 2; John S. Dryzek, *Democracy in Capitalist Times* (New York: Oxford University Press, 1996); Stanley Aronowitz and Peter Bratsis, eds., *Paradigm Lost: State Theory Reconsidered* (Minneapolis: University of Minnesota Press, 2002); Paul Wetherly, *Marxism and the State: An Analytical Approach.* (New York: Palgrave MacMillan, 2005).

67. Harvey Molotch, "The City as a Growth Machine: Towards of Political Economy of Place," *American Journal of Sociology* 82, no. 2 (1976): 309–322; Harvey Molotch, "Capital and Neighborhood in the United States," *Urban Affairs Quarterly* 14, no. 3 (1979): 289–312; Logan and Molotch, *Urban Fortunes.*

68. Weiss, *The Rise of the Community Builders*; Gonzalez, *Urban Sprawl, Global Warming, and the Empire of Capital*, chap. 3.

69. John S. Duffield, *Over a Barrel: The Costs of U.S. Foreign Oil Dependence* (Stanford, CA: Stanford University Press, 2008); International Energy Agency, *CO2 Emissions from Fuel Combustion*, 2010 ed. (Paris: International Energy Agency, 2010); Marilyn A. Brown and Benjamin K. Sovacool, *Climate Change and Global Energy Security: Technology and Policy Options* (Cambridge, MA: MIT Press, 2011) John M. Deutch, *The Crisis in Energy Policy* (Cambridge, MA: Harvard University Press, 2011); Michael J. Graetz, *The End of Energy: The Unmaking of America's Environment, Security, and Independence* (Cambridge, MA: MIT Press, 2011).

Table 2.1. Carbon Dioxide (CO2) Per Capita Emissions of Selected Countries*

Countries	Per Capita CO2 Emissions (in tons)
United States	**18.4**
Russia	11.2
South Korea	10.3
Germany	9.8
Japan	9
United Kingdom	8.3
France	5.7
China	5
India	1.3

Source: International Energy Agency, *CO2 Emissions from Fuel Combustion*, p. 49–51.
*All figures are for 2008 and all selected countries have populations over 35 million

Table 2.2. Energy Consumption Per Capita of Selected Countries*

Countries	Energy Consumption Per Capita (Ton of Equivalent Oil)
United States	**7.8**
Russia	4.7
South Korea	4.5
France	4.4
Japan	4.1
European Union (27)	3.7
China	1.4
India	0.5

Source: Christian de Perthuis, *Economic Choices in a Warming World* (New York: Cambridge University Press, 2011), 238–239.
*All selected countries have populations over 35 million

70. Daniel C. Esty and Michael E. Porter, "Pain at the Pump? We Need More," *New York Times*, Apr. 28, 2011, A25.

71. Vice President Richard "Dick" Cheney, as head of an energy task force, famously rejected conservation in 2001 as part of U.S. energy policy. Joseph Kahn, "Cheney Promotes Increasing Supply as Energy Policy," *New York Times*, May 1, 2001, A1. Also see Peter Calthorpe, *Urbanism in the Age of Climate Change* (Washington, DC: Island Press, 2011).

Chapter 3

1. *Rockefeller Foundation Annual Report, 1956* (New York: Rockefeller Foundation, 1957), 11–12.

2. Peter Collier and David Horowitz, *The Rockefellers: An American Dynasty* (New York: Holt, Rinehart, and Winston, 1976); Raymond B. Fosdick, *The Story of the Rockefeller Foundation* (New York: Transaction Publishers, 1988 [1952]); Peter J. Johnson and John Ensor Harr, *The Rockefeller Century: Three Generations of America's Greatest Family* (New York: Scribner, 1988).

3. Steven Palmer, *Launching Global Health: The Caribbean Odyssey of the Rockefeller Foundation* (Ann Arbor: University of Michigan Press, 2010); Nicolas Guilhot, ed., *The Invention of International Relations Theory: Realism, the Rockefeller Foundation, and the 1954 Conference on Theory* (New York: Columbia University Press, 2011).

4. Edward H. Berman, *Influence of the Carnegie, Ford, and Rockefeller Foundations on American Foreign Policy* (Albany: State University of New York Press, 1983), 5.

5. Richard G. Hewlett and Oscar E. Anderson, Jr., *The New World, 1939/1946: A History of the United States Atomic Energy Commission* (University

Park: Pennsylvania State University Press, 1962); Corbin Allardice and Edward R. Trapnell, *The Atomic Energy Commission* (New York: Praeger, 1974), 8, 11; G. Pascal Zachary, *Endless Frontier: Vannevar Bush: Engineer of the American Century* (New York: Free Press, 1997).

6. Zachary, *Endless Frontier*, 82.

7. Zachary, *Endless Frontier*, 83–84.

8. Allardice and Trapnell, *The Atomic Energy Commission*, chap. 5.

9. Richard G. Hewlett, and Francis Duncan, *Atomic Shield, 1947/1952: A History of the United States Atomic Energy Commission* (Pennsylvania State University Press, 1969), 115–117.

10. Allardice and Trapnell, *The Atomic Energy Commission*, 62. Also see Richard T. Sylves, *The Nuclear Oracles: A Political History of the General Advisory Committee of the Atomic Energy Commission, 1947–1977* (Ames: Iowa State University Press, 1987).

11. As quoted in Daniel Ford, *The Cult of the Atom: The Secret Papers of the Atomic Energy Commission* (New York: Simon and Schuster, 1982), 33. Also see Allardice and Trapnell, *The Atomic Energy Commission*, 104; and Richard Pfau, *No Sacrifice Too Great: The Life of Lewis L. Strauss* (Charlottesville: University Press of Virginia, 1984), 99.

12. Pfau, *No Sacrifice Too Great*, 38.

13. Pfau, *No Sacrifice Too Great*, 84.

14. Pfau, *No Sacrifice Too Great*, 54.

15. Allardice and Trapnell, *The Atomic Energy Commission*, 33–34.

16. Pfau, *No Sacrifice Too Great*, 128.

17. Pfau, *No Sacrifice Too Great*, 185–186.

18. Pfau, *No Sacrifice Too Great*, 211.

19. Richard G. Hewlett and Jack M. Holl, *Atoms for Peace and War 1953–1961: Eisenhower and the Atomic Energy Commission* (Berkeley: University of California Press, 1989), 13.

20. Hewlett and Holl, *Atoms for Peace and War*, 194.

21. Panel on the Impact of the Peaceful Uses of Atomic Energy, *Peaceful Uses of Atomic Energy*, vol. 1 (Washington, DC: Government Printing Office, 1956), xi.

22. Panel on the Impact of the Peaceful Uses of Atomic Energy, *Peaceful Uses of Atomic Energy*, vol. 2 (Washington, DC: Government Printing Office, 1956), xiii–xviii.

23. Panel on the Impact of the Peaceful Uses of Atomic Energy, *Peaceful Uses of Atomic Energy*, vol. 1, p. 2, emphasis added.

24. Panel on the Impact of the Peaceful Uses of Atomic Energy, *Peaceful Uses of Atomic Energy*, vol. 1, p. 95.

25. Panel on the Impact of the Peaceful Uses of Atomic Energy, *Peaceful Uses of Atomic Energy*, vol. 1, p. 97.

26. Everett L. Hollis, "The United States Atomic Energy Act of 1954—A Brief Survey," in *The Economics of Nuclear Power: Including Administration and Law*, J. Guéron, J. A. Lane, I. R. Maxwell, and J. R. Menke, eds. (New York: McGraw-Hill, 1957), 495–496.

27. Mark Hertsgaard, *Nuclear Inc.: The Men and Money behind Nuclear Energy* (New York: Pantheon, 1983), 282–287; David Nye, *Image Worlds: Corporate Identities at General Electric, 1890-1930* (Cambridge, MA: MIT Press, 1985).

28. Frank T. Kryza, *The Power of Light: The Epic Story of Man's Quest to Harness the Sun* (New York: McGraw-Hill, 2003), 22.

29. Kryza, *The Power of Light*, 73.

30. Kryza, *The Power of Light*, 75.

31. Kryza, *The Power of Light*, 26–27.

32. Kryza, *The Power of Light*, 27.

33. Kryza, *The Power of Light*, 141–146.

34. Kryza, *The Power of Light*, 259. Also see, Alexis Madrigal, *Powering the Dream: The History and Promise of Green Technology* (Cambridge, MA: Da Capo Press, 2011), chap. 14.

35. *Proceedings of the World Symposium on Applied Solar Energy, Phoenix, AZ, Nov. 1–5, 1955* (San Francisco: Jorgenson & Co., 1956), 3.

36. Harvey Strum, "The Association for Applied Solar Energy/Solar Energy Society, 1954–1970," *Technology and Culture* 26, no. 3 (1985): 572.

37. *Proceedings of the World Symposium on Applied Solar Energy*, 15–16.

38. *Proceedings of the World Symposium on Applied Solar Energy*, 303.

39. Henry B. Sargent, "The Association for Applied Solar Energy," in *Proceedings of the World Symposium on Applied Solar Energy*, 18.

40. Harvey Strum and Fred Strum, "American Solar Energy Policy, 1952–1970," *Environmental Review* 7 (summer 1983): 136.

41. Strum, "Association for Applied Solar Energy/Solar," 578.

42. Frank G. Dawson, *Nuclear Power: Development and Management of a Technology* (Seattle: University of Washington Press, 1976), 77.

Chapter 4

1. Marc Weiss, *The Rise of the Community Builders: The American Real Estate Industry and Urban Land Planning* (New York: Columbia University Press, 1987); George A. Gonzalez, *The Politics of Air Pollution: Urban Growth, Ecological Modernization, and Symbolic Inclusion* (Albany: State University of New York Press, 2005), chap. 4; and "Urban Sprawl, Global Warming, and the Limits of Ecological Modernization." *Environmental Politics* 14, no. 3 (2005): 34–362.

2. Robert Fishman, *Bourgeois Utopias: The Rise and Fall of Suburbia* (New York: Basic Books, 1987); Lizabeth Cohen, "Is There an Urban History of Consumption?," *Journal of Urban History* 29, no. 2 (2003): 87–106; Robert M. Fogelson, *Bourgeois Nightmares: Suburbia, 1870–1930* (New Haven, CT: Yale University Press, 2005); Paul L. Knox, *Metroburbia, USA* (Piscataway, NJ: Rutgers University Press, 2008).

3. *Report of the President's Conference on Unemployment* (Washington, DC: Government Printing Office, 1921), 7–14.

4. *Report of the President's Conference on Unemployment*, 20.

5. *Report of the President's Conference on Unemployment*, 9, 14, 89.

6. *Report of the President's Conference on Unemployment*, 96, emphasis added.

7. Frederic L. Paxson, "The American Highway Movement, 1916–1935," *American Historical Review* 51, no. 2 (1946): 239–241.

8. American Road Congress, *Papers, Addresses, and Resolutions before the American Road Congress, Richmond, Virginia, November 20–23, 1911* (Baltimore: Waverly Press, 1911).

9. Hugh Chalmers, "Relation of the Automobile Industry to the Good Roads Movement, in American Road Congress," in *Papers, Addresses, and Resolutions before the American Road Congress, Richmond, Virginia, November 20–23, 1911* (Baltimore: Waverly Press, 1911), 142–143.

10. Chalmers, "Relation of the Automobile Industry to the Good Roads Movement," 149.

11. *Report of the President's Conference on Unemployment*, 21.

12. *Report of the President's Conference on Unemployment*, 118.

13. Marc Weiss, *The Rise of the Community Builders: The American Real Estate Industry and Urban Land Planning* (New York: Columbia University Press, 1987), 67.

14. Adam Rome, *The Bulldozer in the Countryside: Suburban Sprawl and the Rise of American Environmentalism* (Cambridge: Cambridge University Press, 2001), chap. 1; also see Gail Radford, *Modern Housing for America: Policy Struggles in the New Deal Era* (Chicago: University of Chicago Press, 1996).

15. Rome, *Bulldozer in the Countryside*, 22–23.

16. *Report of the President's Conference on Unemployment*, 118.

17. Greg Hise, *Magnetic Los Angeles: Planning the Twentieth-Century Metropolis* (Baltimore: Johns Hopkins University Press, 1997), 38.

18. Weiss, *Rise of the Community Builders*, 29.

19. Peter Fearon, *War, Prosperity, and Depression: The U.S. Economy 1917–45* (Lawrence: University Press of Kansas, 1987), 48

20. Ann Markusen, *Profit Cycles, Oligopoly, and Regional Development* (Cambridge: MIT Press, 1985); Robert D. Atkinson, *The Past and Future of America's Economy: Long Waves of Innovation That Power Cycles of Growth* (Northampton, MA: Edward Elgar, 2004).

21. David A. Hounshell, *From the American System to Mass Production, 1800–1932: The Development of Manufacturing Technology in the United States* (Baltimore: Johns Hopkins University Press, 1984), chaps. 6, 7, 8.

22. Alexander J. Field, "Technological Change and U.S. Productivity Growth in the Interwar Years," *Journal of Economic History* 66, no. 1 (2006): 206.

23. Fearon, *War, Prosperity, and Depression*, 55.

In the following, historian T. C. Barker reports automobile ownership during the 1930s among the leading economies of the world at the time:

> There were then [1939] only 2,000,000 cars of all makes registered in the whole country [of Great Britain] (and 460,000 motor cycles), while the United States, with less than three times the population, possessed 30,000,000 cars. And Britain was well ahead of the other Europeans. France, for instance, had only 1,600,000 cars in 1938 and Germany, still at an earlier stage of market growth, had fewer: 1,100,000 cars (and 1,300,000 motor cycles). T. C. Barker, "The International History of Motor Transport," *Journal of Contemporary History* 20, no. 1 (1985): 3–19.

24. Fearon, *War, Prosperity, and Depression*, 55.

25. Fearon, *War, Prosperity, and Depression*, 58; also see Jean-Pierre Bardou, Jean-Jacques Chanaron, Patrick Fridenson, and James M. Laux, *The Automobile Revolution: The Impact of an Industry* (Chapel Hill: University of North Carolina Press, 1982); David J. St. Clair, *The Motorization of American Cities* (New York: Praeger, 1986); Matthew Paterson, *Automobile Politics* (New York: Cambridge University Press, 2007).

26. Elliot Rosen, *Roosevelt, the Great Depression, and the Economics of Recovery* (Charlottesville: University of Virginia Press, 2005), 118.

27. Maury Klein, *The Genesis of Industrial America, 1870–1920* (New York: Cambridge University Press, 2007), 181.

28. Richard B. Du Boff, *Accumulation and Power: An Economic History of the United States* (Armonk, NY: M.E. Sharpe, 1989), 83.

29. Josephine Young Case, *Owen D. Young and American Enterprise: A Biography* (Boston: David R. Godine, 1982); David Nye, *Image Worlds: Corporate Identities at General Electric, 1890-1930* (Cambridge, MA: MIT Press, 1985).

30. Report of the Committee on Recent Economic Changes, of the President's Conference on Unemployment, *Recent Economic Changes in the United States*, vols. 1–2 (New York: McGraw-Hill, 1929), v.

31. Ibid.

32. Report of the Committee on Recent Economic Changes, of the President's Conference on Unemployment, *Recent Economic Changes in the United States*, 236.

33. Report of the Committee on Recent Economic Changes, of the President's Conference on Unemployment, *Recent Economic Changes in the United States*, 254.

34. Report of the Committee on Recent Economic Changes, of the President's Conference on Unemployment, *Recent Economic Changes in the United States*, 422.

35. Martha L. Olney, *Buy Now, Pay Later: Advertising, Credit, and Consumer Durables in the 1920s* (Chapel Hill: University of North Carolina Press, 1991).

36. Sidney Hyman, *Marringer S. Eccles: Private Entrepreneur and Public Servant* (Stanford, CA: Stanford University Graduate School of Business, 1976), 144.

37. Hyman, *Marringer S. Eccles*, 142; also see Rudy Abramson, *Spanning the Century: The Life of W. Averell Harriman, 1891–1986* (New York: W. Morrow, 1992).

38. Hyman, *Marringer S. Eccles*, 142.

39. Hyman, *Marringer S. Eccles*, 142. This board headed up the Federal Home Loan Bank System, created in 1932. It was made up of eleven regionally based home loan banks that served as a central credit agency, similar to that of the Federal Reserve System.

40. Paxson, "The American Highway Movement, 1916–1935," 242; Gonzalez, *Politics of Air Pollution*, chap. 4.

41. James Dunn, *Driving Forces: The Automobile, Its Enemies, and the Politics of Mobility* (Washington, DC: Brookings Institution Press, 1998).

42. Jane Holtz Kay, *Asphalt Nation: How the Automobile Took Over America and How We Can Take It Back* (Berkeley: University of California Press, 1998), 205.

43. Paxson, "The American Highway Movement, 1916–1935," 250.

44. Stan Luger, *Corporate Power, American Democracy, and the Automobile Industry* (New York: Cambridge University Press, 2000); Stan Luger, "Review of *Sloan Rules: Alfred P. Sloan and the Triumph of General Motors*," *American Historical Review* 110, no. 1 (2005): 174; also see Kay, *Asphalt Nation*, 218–219.

45. Bradford Snell, *American Ground Transport* (Washington, DC: U.S. Government Printing Office, 1974); Glen Yago, *The Decline of Transit: Urban Transportation in German and U.S. Cities, 1900–1970* (New York: Cambridge University Press, 1984), chap. 4; Scott Bottles, *Los Angeles and the Automobile: The Making of the Modern City* (Los Angeles: University of California Press, 1987).

46. Hyman, *Marringer S. Eccles*, 141.

47. Hyman, *Marringer S. Eccles*, 143.

48. Robert Fishman, *Bourgeois Utopias: The Rise and Fall of Suburbia* (New York: Basic Books, 1987); Marc Weiss, *The Rise of the Community*

Builders: The American Real Estate Industry and Urban Land Planning (New York: Columbia University Press, 1987); John Stilgoe, *Borderland: Origins of the American Suburb, 1820–1939* (New Haven, CT: Yale University Press, 1988); Robert Bruegmann, *Sprawl: A Compact History* (Chicago: University of Chicago Press, 2005); Robert M. Fogelson, *Bourgeois Nightmares: Suburbia, 1870–1930* (New Haven, CT: Yale University Press, 2005).

49. Weiss, *The Rise of the Community Builders*, 146.

50. Julian H. Zimmerman, *The FHA Story in Summary, 1934–1959* (Washington, DC: U.S. Federal Housing Administration, 1959), 7–8.

51. Jeffrey M. Hornstein, *A Nation of Realtors: A Cultural History of the Twentieth-Century American Middle Class* (Durham, NC: Duke University Press, 2005), 150.

52. Weiss, *The Rise of the Community Builders*, 146; Binyamin Appelbaum, "Without Loan Giants, 30-Year Mortgage May Fade Away," *New York Times*, Mar. 4, 2011, A1.

53. Weiss, *The Rise of the Community Builders*, 146.

54. Weiss, *The Rise of the Community Builders*, 148.

55. Weiss, *The Rise of the Community Builders*, 147; also see Hornstein, *A Nation of Realtors*, 150–152.

56. Kay, *Asphalt Nation*, 201.

57. Peter O. Muller, *Contemporary Suburban America* (Englewood Cliffs, NJ: Prentice-Hall, 1981), 44.

58. Kay, *Asphalt Nation*, 201.

59. Kenneth T. Jackson, *Crabgrass Frontier: The Suburbanization of the United States* (New York: Oxford University Press, 1985), 206.

60. Jackson, *Crabgrass Frontier*, 209.

By promoting low-density urban development and sprawl, the federal government eschewed a network of intellectuals advocating what was known as *social housing* during the 1920s and 1930s. Drawing from experiences in Europe, such thinkers argued that apartment complex housing that emphasized community living—wherein services like day care and schooling for children were provided, as well as recreational facilities and activities—were economically and socially preferable to suburban tract housing. Historian Gail Radford points out that the few U.S. experiments in social housing proved to be successful. These projects, built in the 1930s—in places like Philadelphia and Harlem—were well planned, affordable, aesthetically pleasing, and provided important amenities to its residents. Moreover, Radford finds that residents of U.S. social housing generally found living there to be agreeable and advantageous. Whereas federal housing projects built in the post–World War II period only allowed the poor, the housing reformers' experiments were mostly occupied by white and blue collar workers. Gail Radford, *Modern Housing for America: Policy Struggles in the New Deal Era* (Chicago: University of Chicago Press, 1996); Nicholas Bloom, *Public Housing*

That Worked: New York in the Twentieth Century (Philadelphia: University of Pennsylvania Press, 2008).

61. Martha L. Olney, *Buy Now, Pay Later: Advertising, Credit, and Consumer Durables in the 1920s* (Chapel Hill: University of North Carolina Press, 1991).

62. Michael French, *U.S. Economic History since 1945* (Manchester: Manchester University Press, 1997); Robert Brenner, *The Boom and the Bubble: The U.S. in the World Economy* (New York: Verso, 2002); Robert Brenner, "New Boom or New Bubble: The Trajectory of the U.S. Economy," *New Left Review* 25 (Jan./Feb. 2004), 57–102; Norman Frumkin, *Tracking America's Economy* (Armonk, NY: M.E. Sharpe, 2004); Louis Uchitelle, "Goodbye, Production (and Maybe Innovation)," *New York Times*, Dec. 24, 2006, sec. 3, p. 4.

63. Peter S. Goodman, "The Economy: Trying to Guess What Happens Next," *New York Times*, Nov. 25, 2007, sec. 4, p. 1.

64. Martin Fackler, "Toyota Expects Decline in Annual Profit," *New York Times*, May 9, 2008, C3.

65. Keith Bradsher, *High and Mighty: SUVs—The World's Most Dangerous Vehicles and How They Got That Way* (New York: Public Affairs, 2002); John A. C. Conybeare, *Merging Traffic: The Consolidation of the International Automobile Industry* (Lanham, MD: Rowman & Littlefield, 2004); Helmut Becker, *High Noon in the Automotive Industry* (New York: Springer, 2006), 12; Neal E. Boudette and Norihiko Shirouzu, "Car Makers' Boom Years Now Look Like a Bubble," *Wall Street Journal*, May 20, 2008, A1.

66. Todd Zaun, "Honda Tries to Spruce Up a Stodgy Image," *New York Times*, Mar. 19, 2005, C3; Martin Fackler, "Toyota's Profit Soars, Helped by U.S. Sales," *New York Times*, Aug. 5, 2006, C4.

67. Nick Bunkley, "Toyota Ahead of G.M. in 2008 Sales," *New York Times*, Jan. 22, 2009, B2.

Chapter 5

1. Frank N. Laird, *Solar Energy, Technology Policy, and Institutional Values* (New York: Cambridge University Press, 2001).

2. The *New York Times* in 2011 reported that "American steel makers never developed the equipment needed for the next generation of nuclear plants." Mathew L. Wald, "Nuclear Industry Thrives in the U.S., but for Export," *New York Times*, Mar. 30, 2011, F2.

3. Francisco Parra, *Oil Politics: A Modern History of Petroleum*. New York: I.B. Tauris, 2004); Harvey Blatt, *America's Environmental Report Card: Are We Making the Grade?* (Cambridge, MA: MIT Press, 2005), 100; and *America's Environmental Report Card: Are We Making the Grade?*, 2nd ed. (Cambridge, MA: MIT Press, 2011), 140; Roy L. Nersesian, *Energy for the*

21st Century (Armonk, NY: M.E. Sharpe, 2007), 205; John S. Duffield, *Over a Barrel: The Costs of U.S. Foreign Oil Dependence* (Stanford, CA: Stanford University Press, 2008); Steffen Hertog, *Princes, Brokers, and Bureaucrats: Oil and the State in Saudi Arabia* (Ithaca, NY: Cornell University Press, 2010); Steve Yetiv, *Explaining Foreign Policy: U.S. Decision-Making in the Gulf Wars* (Baltimore: Johns Hopkins University Press, 2011).

4. Peter O. Muller, *Contemporary Suburban America* (Englewood Cliffs, NJ: Prentice-Hall, 1981); Robert A. Beauregard, *When America Became Suburban* (Minneapolis: University of Minnesota Press, 2006); Paul L. Knox, *Metroburbia, USA* (Piscataway, NJ: Rutgers University Press, 2008).

5. Mark S. Foster, *From Streetcar to Superhighway: American City Planners and Urban Transportation, 1900–1940* (Philadelphia: Temple University Press, 1981); Jeffrey R. Kenworthy and Felix B. Laube, with Peter Newman, Paul Barter, Tamim Raad, Chamlong Poboon, and Benedicto Guia, Jr., *An International Sourcebook of Automobile Dependence in Cities 1960–1990* (Boulder: University Press of Colorado, 1999).

6. John M. Blair, *The Control of Oil* (New York: Pantheon, 1976); Ed Shaffer, *The United States and the Control of World Oil* (New York: St. Martin's, 1983); George Philip, *The Political Economy of International Oil* (Edinburgh: Edinburgh University Press, 1994).

7. American Petroleum Institute, *Petroleum Facts and Figures: Centennial Edition* (New York: American Petroleum Institute, 1959), 246–247.

8. Daniel Yergin, *The Prize: The Epic Quest for Oil, Money, and Power* (New York: Simon & Schuster, 1991); Kenneth S. Deffeyes, *Hubbert's Peak: The Impending World Oil Shortage* (Princeton, NJ: Princeton University Press, 2001).

9. Blair, *The Control of Oil*; Ed Shaffer, *The United States and the Control of World Oil*; Richard H. Vietor, *Energy Policy in America since 1945* (New York: Cambridge University Press, 1984); Philip, *The Political Economy of International Oil*; Ian Rutledge, *Addicted to Oil: America's Relentless Drive for Energy Security* (New York: I. B. Tauris, 2005); Rachel Bronson, *Thicker Than Oil: America's Uneasy Partnership with Saudi Arabia* (New York: Oxford University Press, 2006).

10. James A. Bill, *The Eagle and the Lion: The Tragedy of American-Iranian Relations* (New Haven, CT: Yale University Press, 1988); John G. Ikenberry, *Reasons of State: Oil Politics and the Capacities of American Government* (Ithaca, NY: Cornell University Press, 1988); Simon Bromley, *American Hegemony and World Oil: The Industry, the State System and the World Economy* (University Park: Pennsylvania State University Press, 1991); Steve A. Yetiv, *Crude Awakenings: Global Oil Security and American Foreign Policy* (Ithaca, NY: Cornell University Press, 2004); Robert J. Pauly, Jr., *U.S. Foreign Policy and the Persian Gulf: Safeguarding American Interests through Selective Multilateralism* (Burlington, VT: Ashgate, 2005); Andrew Scott Cooper, *The Oil*

Kings: How the U.S., Iran, and Saudi Arabia Changed the Balance of Power in the Middle East (New York: Simon & Schuster, 2011).

11. E.g., Frederic Dewhurst, and the Twentieth Century Fund, *America's Needs and Resources: A New Survey* (New York: Twentieth Century Fund, 1955); Thomas Reynolds Carskadon and George Henry Soule, *USA in New Dimensions: The Measure and Promise of America's Resources, A Twentieth Century Fund Survey* (New York: Macmillan, 1957); Arnold B. Barach, and the Twentieth Century Fund, *USA and Its Economic Future: A Twentieth Century Fund Survey* (New York: Macmillan, 1964).

12. "As Oil Consultant, He's without Like or Equal," *New York Times,* July 27, 1969, sec. 3, p. 3; Shaffer, *The United States and the Control of World Oil,* 214–218.

In a 1969 profile of Walter J. Levy titled "As Oil Consultant, He's without Like or Equal," the *New York Times* noted that "he is readily acknowledged as the 'dean of oil consultants' even by competitors." The profile went on to explain that "there are few, if any, major oil controversies in which Mr. Levy has not acted as a consultant," and that he "has been an advisor to most of the major oil companies, most of the important consuming countries and many of the large producing countries." "As Oil Consultant, He's without Like or Equal."

13. Walter L. Buenger and Joseph A. Pratt, *But Also Good Business: Texas Commerce Banks and the Financing of Houston and Texas, 1886–1986* (College Station: Texas A&M University Press, 1986), 299.

14. Twentieth Century Fund Task Force on the International Oil Crisis, *Paying for Energy* (New York: McGraw-Hill, 1975), vii–viii; Twentieth Century Fund Task Force on United States Energy Policy, *Providing for Energy* (New York: McGraw-Hill, 1977), xi–xii; Richard Magat, *The Ford Foundation at Work: Philanthropic Choices, Methods, and Styles* (New York: Plenum, 1979), 165, 172; Robin W. Winks, *Laurence S. Rockefeller: Catalyst for Conservation* (Washington, DC: Island Press, 1997), 44, 196; G. Pascal Zachary, *Endless Frontier: Vannevar Bush: Engineer of the American Century* (New York: Free Press, 1997), 83–84.

15. Twentieth Century Fund Task Force on the International Oil Crisis, *Paying for Energy,* 9.

16. Twentieth Century Fund Task Force on the International Oil Crisis, *Paying for Energy,* 9, emphasis in original.

17. Twentieth Century Fund Task Force on United States Energy Policy, *Providing for Energy,* 5, emphasis in original.

18. Ibid.

19. Twentieth Century Fund Task Force on United States Energy Policy, *Providing for Energy,* 23, emphasis in original.

20. Twentieth Century Fund Task Force on United States Energy Policy, *Providing for Energy,* 23–24.

21. Twentieth Century Fund Task Force on United States Energy Policy, *Providing for Energy,* 24.

22. Twentieth Century Fund Task Force on the International Oil Crisis, *Paying for Energy*, 15.

23. "Energy Efficiency Fails to Cut Consumption—Study," Reuters, Nov. 27, 2007; Horace Herring and Steve Sorrell, eds., *Energy Efficiency and Sustainable Consumption: The Rebound Effect* (New York: Palgrave Macmillan, 2009).

24. Winston Harrington and Virginia McConnell, *Resources for the Future Report: Motor Vehicles and the Environment* (Washington, DC: Resources for the Future, 2003), chaps. 6, 7.

25. Energy Information Administration, *Annual Energy Review 2010* (Washington, DC: U.S. Department of Energy, 2011), 59.

26. Rutledge, *Addicted to Oil*, 10.

27. The U.S. Energy Department reports that in 2008, about 8.7 million barrels of oil per day (mb/d) were used to power the U.S. automobile fleet (including light trucks and motorcycles). Global petroleum production in 2008 was 85.8 mb/d. In spite of the recession of 2009, the amount of oil used to power the U.S. automotive fleet remained at 8.7 mb/d. Stacy C. Davis, Susan W. Diegel, and Robert G. Boundy, *Transportation Energy Data Book*, 29th ed. (Washington, DC: Department of Energy, 2010), table 1.4, chap. 1, p. 5, table 1.16, chap. 1, p. 22; Stacy C. Davis, Susan W. Diegel, and Robert G. Boundy, *Transportation Energy Data Book*, 30th ed. (Washington, DC: Department of Energy, 2011), table 1.15, chap. 1, p. 21.

28. Duffield, *Over a Barrel*, chap. 2.

29. In 2010, the U.S. Department of Energy noted that "the United States has accounted for almost one-quarter of the world's petroleum consumption for the last two decades, but in 2009 accounted for only 22.2 percent." American oil consumption in 2007 was 20.7 million barrels per day (mb/d)—24 percent of total world production (86.1 mb/d). The economic recession that began in 2008 resulted in a drop of U.S. petroleum use to 19.5 mb/d in 2008 and 18.7 mb/d in 2009. Davis, Diegel, and Boundy, *Transportation Energy Data Book*, 29th ed., chap. 1, p. 5.

By way of comparison, the International Energy Agency reported that oil consumption in the European Union in 2009 was 12.2 mb/d—14.5 percent of total world petroleum production (84 mb/d). The European Union in 2009 had a population of 500 million, whereas the United States had one of 300 million. (In 2008, European Union oil consumption was 12.4 mb/d.) China, with about 20 percent of the global population, in 2009 consumed 8.1 mb/d—9.6 percent of world total production. (China, in 2008, used 7.7 mb/d.) (International Energy Agency petroleum figures exclude biofuels, which amounted to 0.8 mb/d of demand in 2008 and 1.1 mb/d in 2009.) International Energy Agency, *World Energy Outlook 2009* (Paris: International Energy Agency, 2009), 81; International Energy Agency, *World Energy Outlook 2010* (Paris: International Energy Agency, 2010), 105. The

New York Times reports that on a per capita basis, U.S. consumption of oil in 2009 was 22 barrels per person. In China it was 2.4 barrels per person. Jad Mouawad, "China's Growth Shifts the Geopolitics of Oil," *New York Times*, Mar. 19, 2010, B1.

30. Philip, *The Political Economy of International Oil*, 195; Paul Roberts, *The End of Oil: On the Edge of a Perilous New World* (New York: Houghton Mifflin, 2004); Rutledge, *Addicted to Oil*, chap. 1; Matthew L. Wald, "When It Comes to Replacing Oil Imports, Nuclear Is No Easy Option, Experts Say," *New York Times*, May 9, 2005, A14; Bruce Podobnik, *Global Energy Shifts: Fostering Sustainability in a Turbulent Age* (Philadelphia: Temple University Press, 2006), chaps. 5, 6; Duffield, *Over a Barrel*, chap. 2.

31. Twentieth Century Fund Task Force on United States Energy Policy, *Providing for Energy*, 24–25, emphasis in original.

32. Jon Van Til, *Living with Energy Shortfall* (Boulder, CO: Westview, 1982); Podobnik, *Global Energy Shifts*, chap. 7; Robert B. Semple, Jr., "Oil and Gas Had Help. Why Not Renewables?" *New York Times*, Oct. 16, 2011, SR10.

33. Kenworthy and Laube, *An International Sourcebook of Automobile Dependence in Cities 1960–1990*; Elisabeth Rosenthal, "Across Europe, Irking Drivers Is Urban Policy," *New York Times*, June 27, 2011, A1.

34. Louis Armand, *Some Aspects of the European Energy Problem: Suggestions for Collective Action* (Paris: Organization for European Cooperation, 1955).

35. Commission for Energy, *Europe's Growing Needs of Energy: How Can They Be Met?* (Paris: Organisation for European Economic Co-Operation, 1956), 25.

36. Commission for Energy, *Europe's Growing Needs of Energy*, 73.

37. Commission for Energy, *Europe's Growing Needs of Energy*, 26.

38. Commission for Energy, *Europe's Growing Needs of Energy*, 73.

39. Commission for Energy, *Europe's Growing Needs of Energy*, 56.

40. Armand, *Some Aspects of the European Energy Problem*, 46.

41. Energy Advisory Commission, *Towards a New Energy Pattern in Europe* (Paris: Organisation for European Economic Co-operation, 1960), 13–14.

42. Energy Advisory Commission, *Towards a New Energy Pattern in Europe*, 83.

43. Energy Advisory Commission, *Towards a New Energy Pattern in Europe*, 61.

44. Energy Advisory Commission, *Towards a New Energy Pattern in Europe*, 83–84.

45. Shaffer, *The United States and the Control of World Oil*, chap. 7; Torleif Haugland, Helge Ole Bergensen, and Kjell Roland, *Energy Structures and Environmental Futures* (New York: Oxford University Press, 1998), 55.

46. Haugland et al., *Energy Structures and Environmental Futures*, 33; also see James Dunn, *Miles to Go: European and American Transportation Policies* (Cambridge, MA: MIT Press, 1981); Nigel Lucas, *Western European Energy Policies: A Comparative Study of the Influence of Institutional Structures on Technical Change* (Oxford: Clarendon, 1985).

47. Harvey Blatt, *America's Environmental Report Card: Are We Making the Grade?*, 2nd ed. (Cambridge, MA: MIT Press, 2011), 142–143.

48. Simon Romero, "Oil-Rich Norwegians Take World's Highest Gasoline Prices in Stride," *New York Times*, Apr. 30, 2005, C1; also see Molly O'Meara Sheehan, *City Limits: Putting the Brakes on Sprawl* (Washington, DC: Worldwatch Institute, 2001).

49. Laird, *Solar Energy, Technology Policy, and Institutional Values*; Travis Bradford, *Solar Revolution: The Economic Transformation of the Global Energy Industry* (Cambridge, MA: MIT Press, 2006).

50. Dorothy Nelkin and Michael Pollak, *The Atom Besieged: Antinuclear Movements in France and Germany* (Cambridge, MA: MIT Press, 1981).

51. Irvin C. Bupp and Jean-Claude Derian, *The Failed Promise of Nuclear Power: The Story of Light Water* (New York: Basic Books, 1978); Peter Stoett, "Toward Renewed Legitimacy? Nuclear Power, Global Warming, and Security," *Global Environmental Politics* 3, no. 1 (2003): 99–116; Jane Dawson and Robert Darst, "Meeting the Challenge of Permanent Nuclear Waste Disposal in an Expanding Europe: Transparency, Trust and Democracy," *Environmental Politics* 15, no. 4 (2006): 610–627; Robert Vandenbosch and Susanne E. Vandenbosch, *Nuclear Waste Stalemate: Political and Scientific Controversies* (Salt Lake City: University of Utah Press, 2007); Max S. Power, *America's Nuclear Wastelands: Politics, Accountability, and Cleanup* (Pullman: Washington State University Press, 2008); Matthen L. Wald, "As Nuclear Waste Languishes, Expense to U.S. Rises," *New York Times*, Feb. 17, 2008, A22; Richard Burleson Stewart and Jane Bloom Stewart, *Fuel Cycle to Nowhere: U.S. Law and Policy on Nuclear Waste* (Nashville, TN: Vanderbilt University Press, 2011); Matthew L. Wald, "A Safer Nuclear Crypt," *New York Times*, July 6, 2011, B1.

52. Michael T. Hatch, *Politics and Nuclear Power: Energy Policy in Western Europe* (Lexington: University Press of Kentucky, 1986).

53. James M. Jasper, *Nuclear Politics: Energy and the State in the United States, Sweden, and France* (Princeton, NJ: Princeton University Press, 1990); Steven Erlanger, "French Plans For Energy Reaffirm Nuclear Path," *New York Times*, Aug. 17, 2008, A6; Gabrielle Hecht, *The Radiance of France: Nuclear Power and National Identity after World War II* (Cambridge, MA: MIT Press, 2009).

54. Henry Nau, *National Politics and International Technology: Nuclear Reactor Development in Western Europe* (Baltimore: Johns Hopkins University Press, 1974); James Kanter, "German Chancellor Calls for Tests of Europe's Nuclear Reactors," *New York Times*, Mar. 24, 2011, B3.

55. Between 1981 and 1986, U.S. daily consumption of petroleum increased by 120,000 barrels, whereas Western Europe consumption dropped by 490,000 barrels. Philip, *The Political Economy of International Oil*, 195; also see Hatch, *Politics and Nuclear Power*; Peter Nijkamp, *Sustainable Cities in Europe: A Comparative Analysis of Urban Energy-Environmental Policies* (London: Earthscan, 1994); Frank J. Convery, ed., *A Guide to Policies for Energy Conservation: The European Experience* (Northampton, MA: Edward Elgar, 1998); Haugland et al., *Energy Structures and Environmental Futures*; Peter Newman, Timothy Beatley, and Heather Boyer, *Resilient Cities: Responding to Peak Oil and Climate Change* (Washington, DC: Island Press, 2009).

56. Nelkin and Pollak, *The Atom Besieged*; Hecht, *The Radiance of France*; Blatt, *America's Environmental Report Card*, 2nd ed., 216.

Chapter 6

1. Steve Yetiv, *Crude Awakenings: Global Oil Security and American Foreign Policy* (Ithaca, NY: Cornell University Press, 2004); Doug Stokes and Sam Raphael, *Global Energy Security and American Hegemony* (Baltimore: Johns Hopkins University Press, 2010).

2. James T. Bartis and Lawrence Van Bibber, *Alternative Fuels for Military Applications* (Santa Monica, CA: RAND Corporation, 2011); Tom Zeller, Jr., "Alternative Fuels Don't Benefit the Military, a RAND Report Says," *New York Times*, Jan. 25, 2011, B1.

3. Intergovernmental Panel on Climate Change (IPCC), *Special Report on Renewable Energy Sources and Climate Change Mitigation* (August 2011).

4. Rex J. Zedalis, *The Legal Dimensions of Oil and Gas in Iraq: Current Reality and Future Prospects* (New York: Cambridge University Press, 2009); Tom Hundley, "Iraq's Oil Industry Poised to Re-enter World Stage," *New York Times*, Feb. 15, 2010. Web.

5. Jon Van Til, *Living with Energy Shortfall* (Boulder, CO: Westview, 1982); Richard H. Vietor, *Energy Policy in America since 1945* (New York: Cambridge University Press, 1984); Bruce Podobnik, *Global Energy Shifts: Fostering Sustainability in a Turbulent Age* (Philadelphia: Temple University Press, 2006), chap. 6.

6. Kenneth S. Deffeyes, *Hubbert's Peak: The Impending World Oil Shortage* (Princeton, NJ: Princeton University Press, 2001); David Goodstein, *Out of Gas: The End of the Age of Oil* (New York: Norton, 2004); Clifford Krauss, "Tapping a Trickle In West Texas," *New York Times*, Nov. 2, 2007, C1.

7. Krauss, "Tapping a Trickle In West Texas"; Ed Crooks and Sheila McNutty, "U.S. Crude Oil Output Highest in a Decade," *Financial Times*, Mar. 3, 2011, A1; Ian Urbina, "Regulators Seek Records on Claims for Gas Wells," *New York Times*, July 30, 2011, A13; Simon Romero, "New Fields

May Propel Americas to Top of Oil Companies' Lists," *New York Times*, Sept. 20, 2011, A1; International Energy Agency, *Oil Market Report* (Paris: International Energy Agency, 2012 Mar. 14), p. 21.

Nersesian notes that Egypt, Indonesia, and the United Kingdom have all reached peak petroleum production, and oil extraction in these countries is in a steady decline. Roy L. Nersesian, *Energy for the 21st Century* (Armonk, NY: M.E. Sharpe, 2007), 199.

Despite the ostensive peaking of petroleum extraction in major producing countries, the U.S. government officially denies the validity of Hubbert's theory. Moreover, prominent officials in the petroleum industry also question its utility. Goodstein, *Out of Gas*; Paul Roberts, *The End of Oil: On the Edge of a Perilous New World* (New York: Houghton Mifflin, 2004); "World Oil Demand to Peak Before Supply—BP," Reuters, Jan. 16, 2008.

8. E.g., Deffeyes, *Hubbert's Peak*; Goodstein, *Out of Gas*; Kenneth S. Deffeyes, *Beyond Oil: The View from Hubbert's Peak* (New York: Hill and Wang, 2005).

9. Nersesian, *Energy for the 21st Century*, 206.

10. Deffeyes, *Hubbert's Peak*, 158. Kenneth S. Deffeyes, *When Oil Peaked* (New York: Hill and Wang, 2010).

11. Stephen Kurczy, "International Energy Agency Says 'Peak Oil' Has Hit," *Christian Science Monitor*, Nov. 11, 2010.

12. Ian Rutledge, *Addicted to Oil: America's Relentless Drive for Energy Security* (New York: I.B. Tauris, 2005), 139.

13. Jad Mouawad and Heather Timmons, "Trading Frenzy Adds to Jump In Price of Oil," *New York Times*, Apr. 29, 2006, A1; Clifford Krauss, "Commodity Prices Tumble," *New York Times*, Oct. 14, 2008, B1.

An alternative argument to explain the recent spike in petroleum prices points to the recent merger of oil companies. This merger trend has resulted in fewer overall oil firms, which facilitates their ability to manipulate prices. Javier Blas, James Boxell, and Kevin Morrison, "Energy Investment Too Small to Meet Growth in Demand, Warns Watchdog," *Financial Times*, May 4, 2005, 1.

14. Paul Roberts, afterward to Paul Roberts, *The End of Oil: On the Edge of a Perilous New World* (New York: Houghton Mifflin, 2005), 335.

15. Matthew R. Simmons, *Twilight in the Desert: The Coming Saudi Oil Shock and the World Economy* (New York: Wiley, 2005), xv, emphasis in original.

16. Nersesian, *Energy for the 21st Century*, 199.

17. Andrés Cala, "Iraq Struggles with High-Flying Oil Goals," *New York Times*, Feb. 22, 2011. Web.

18. Ed Shaffer, *The United States and the Control of World Oil* (New York: St. Martin's Press, 1983); Zbigniew Brzezinski, *The Grand Chessboard: American Primacy and Its Geostrategic Imperatives* (New York: Basic, 1998); David Harvey, *The New Imperialism* (New York: Oxford University Press,

2003); Doug Stokes and Sam Raphael, *Global Energy Security and American Hegemony* (Baltimore: Johns Hopkins University Press, 2010).

19. Greg Palast, *Armed Madhouse* (New York: Dutton, 2006).

In a late 2009 auction most drilling rights to Iraqi oil fields went to non-U.S. oil firms. Hundley, "Iraq's Oil Industry Poised to Re-enter World Stage."

20. In a 2011 *New York Times* article, an executive from a Russian oil company (Lukoil) operating in Iraq was quoted as saying that "the strategic interest of the United States is in new oil supplies arriving on the world market, to lower prices." Andrew E. Kramer, "In Rebuilding Iraq's Oil Industry, U.S. Subcontractors Hold Sway," *New York Times*, June 17, 2011, B1.

21. Stefan Halper and Jonathan Clarke, *America Alone: The Neo-Conservatives and the Global Order* (Cambridge: Cambridge University Press, 2004).

22. National Energy Policy Development Group, *National Energy Policy* (Washington, DC: U.S. Government Printing Office, May 2001), viii.

23. General Accounting Office, *Energy Task Force: Process Used to Develop the National Energy Policy* (Washington, DC: Government Printing Office, 2003), 17–18.

24. National Energy Policy Development Group, *National Energy Policy*, ix.

25. National Energy Policy Development Group, *National Energy Policy*, x.

26. Ibid., xiii.

27. National Energy Policy Development Group, *National Energy Policy*, chap. 8, p. 5.

28. H. Josef Hebert, "Group: Cheney Task Force Eyed on Iraq Oil," Associated Press, July 18, 2003; Terry H. Anderson, *Bush's Wars* (New York: Oxford University Press, 2011), 58.

29. National Energy Policy Development Group, *National Energy Policy*, chap. 8, p. 4, emphasis added.

30. Gary Dorrien, *The Neoconservative Mind: Politics, Culture, and the War of Ideology*, (Philadelphia: Temple University Press, 1993); and *Imperial Designs: Neoconservatism and the New Pax Americana* (New York: Routledge, 2004); John Ehrman, *The Rise of Neoconservatism: Intellectuals and Foreign Affairs, 1945–1994* (New York: Cambridge University Press, 1995); Murray Friedman, *The Neoconservative Revolution: Jewish Intellectuals and the Shaping of Public Policy* (New York: Cambridge University Press, 2005); Benjamin Balint, *Running Commentary: The Contentious Magazine That Transformed the Jewish Left into the Neoconservative Right* (New York: Public Affairs, 2010); Thomas L. Jeffers, *Norman Podhoretz: A Biography* (New York: Cambridge University Press, 2010); C. Bradley Thompson with Yaron Brook, *Neoconservatism: An Obituary for an Idea* (Boulder, CO: Paradigm Publishers, 2010); Justin Vaïsse, *Neoconservatism: The Biography of a Movement* (Cambridge, MA: Harvard University Press, 2010);

Jean-François Drolet, *American Neoconservatism: The Politics and Culture of a Reactionary Idealism* (New York: Columbia University Press, 2011).

31. New American Century, "To The Honorable William J. Clinton, President of the United States," January 26, 1998, http://www.newamericancentury.org/iraqclintonletter.htm (viewed February 25, 2010), emphasis added.

32. Dorrien, *Imperial Designs*, chap. 4.

33. Dorrien, *Imperial Designs*, 143; Halper and Clarke, *America Alone*.

34. Dorrien, *Imperial Designs*, 142.

35. James Mann, *Rise of the Vulcans: The History of Bush's War Cabinet* (New York: Viking, 2004), 369.

36. Mann, *Rise of the Vulcans*, 374.

37. Mann, *Rise of the Vulcans*, 231.

38. The economic elite–led policy-planning network has two groupings—one that is characterized as "moderate" or "corporate liberal" and the other as "conservative." Although these two groups will frequently compromise on issues, they sometimes cannot. When they cannot find common ground, their struggles will usually spill over into government, where each will utilize its political strength to get its way. James Weinstein, *The Corporate Ideal in the Liberal State: 1900–1918* (Boston: Beacon Press, 1968), chap. 1; David Eakins, "Business Planners and America's Postwar Expansion," in David Horowitz, ed., *Corporations and the Cold War* (New York: Monthly Review Press, 1969); and "Policy-Planning for the Establishment," in Ronald Radosh and Murray N. Rothbard, *A New History of Leviathan* (New York: E.P. Dutton & Co., 1972); Clyde W. Barrow, *Critical Theories of the State* (Madison: Wisconsin University Press, 1993), chap. 1; G. William Domhoff, *The Power Elite and the State* (New York: Aldine de Gruyter, 1990), 38–39; and *Who Rules America?*, 6th ed. (New York: McGraw-Hill, 2009), chap. 3.

39. Dorrien, *Imperial Designs*, 130.

The Bradley Foundation was funded by the late Bradley brothers, owners of the Allen-Bradley company. The foundation is recognized as sponsoring almost exclusively politically conservative organizations and projects. According to its mission statement:

> The Bradley brothers were committed to preserving and defending the tradition of free representative government and private enterprise that has enabled the American nation and, in a larger sense, the entire Western world to flourish intellectually and economically. The Bradleys believed that the good society is a free society. The Lynde and Harry Bradley Foundation is likewise devoted to strengthening American democratic capitalism and the institutions, principles, and values that sustain and nurture it. Its programs support limited, competent government; a dynamic marketplace for economic, intellectual, and cultural activity; and

a vigorous defense, at home and abroad, of American ideas and institutions. The Lynde and Harry Bradley Foundation, The Bradley Foundation's Mission, http://www.bradleyfdn.org/foundations_mission.asp (viewed Feb. 25, 2010).

40. Mann, *Rise of the Vulcans*, 243.

41. Mark Maslin, *Global Warming: A Very Short Introduction* (New York: Oxford University Press, 2009).

42. Rüdiger K. W. Wurzel and James Connelly, eds., *The European Union as a Leader in International Climate Change Politics* (New York: Routledge, 2011).

43. Asia-Pacific Partnership on Clean Development and Climate, http://www.asiapacificpartnership.org/english/default.aspx (viewed on Jan. 24, 2011).

44. Jane Perlez, "U.S. to Join China and India in Climate Pact," *New York Times*, July 27, 2005, A8.

45. Asia-Pacific Partnership on Clean Development and Climate, http://www.asiapacificpartnership.org/english/default.aspx (viewed on Jan. 24, 2011).

46. Asia-Pacific Partnership for Clean Development and Climate, Renewable Energy and Distributed Generation Task Force, http://www.asiapacificpartnership.org/english/tf_renewable_energy.aspx (viewed on Jan. 24, 2011).

47. Asia-Pacific Partnership for Clean Development and Climate, Project Roster—Renewable Energy & Distributed Generation Task Force, http://www.asiapacificpartnership.org/english/pr_renewable_energy.aspx (viewed on Jan. 24, 2011).

48. Ibid.

49. Keith Bradsher, "On Clean Energy, China Skirts Rules," *New York Times*, Sept. 9, 2010, A1.

50. John M. Broder, "Director of Policy on Climate Will Leave, Her Goal Unmet," *New York Times*, Jan. 25, 2011, A15.

51. US-China Energy Forum, http://cleanenergyforum.net/index.cfm (viewed Jan. 26, 2011).

52. US-China Clean Energy Forum, About Us, http://cleanenergyforum.net/aboutus.cfm (viewed on Jan. 26, 2011).

53. US-China Energy Forum, http://cleanenergyforum.net/index.cfm (viewed Jan. 26, 2011).

54. US-China Energy Forum, China-US Clean Energy Initiatives (Beijing: US-China Energy Forum, May 26, 2009).

55. Stephen Schmidheiny and Federico Zorraquin, with the World Business Council for Sustainable Development, *Financing Change: The Financial Community, Ecoefficiency, and Sustainable Development* (Cambridge, MA: MIT Press, 1996), xvi–xx.

56. WBCSD (World Business Council for Sustainable Development), About the WBCSD, http://www.wbcsd.org (viewed on Mar. 7, 2010).

57. Albert Weale, *The New Politics of Pollution* (New York: Manchester University Press, 1992); Maarten A. Hajer, *The Politics of Environmental Discourse* (New York: Oxford University Press, 1995); Arthur P. J. Mol, *Globalization and Environmental Reform: The Ecological Modernization of the Global Economy* (Cambridge, MA: MIT Press, 2001); and "Ecological Modernization and the Global Economy," *Global Environmental Politics* 2, no. 2 (2002): 92–115; Richard York and Eugene A. Rosa, "Key Challenges to Ecological Modernization Theory," *Organization & Environment* 16, no. 3 (2003): 273–288; John S. Dryzek, *The Politics of the Earth*, 2nd ed. (New York: Oxford University Press, 2005), chap. 8; George. A. Gonzalez, *The Politics of Air Pollution: Urban Growth, Ecological Modernization, and Symbolic Inclusion* (Albany: State University of New York Press, 2005); Michael T. Hatch, ed., *Environmental Policymaking: Assessing the Use of Alternative Policy Instruments* (Albany: State University of New York Press, 2005); Arthur P. J. Mol, David A. Sonnenfeld, and Gert Spaargaren, eds., *The Ecological Modernisation Reader: Environmental Reform in Theory and Practice* (London: Routledge, 2009).

58. WBCSD, About the WBCSD, http://www.wbcsd.org (viewed on Mar. 7, 2010).

59. Ibid.

60. Schmidheiny et al., *Financing Change*, xxiv.

61. Peter Christoff, "Ecological Modernization, Ecological Modernities," *Environmental Politics* 5, no. 3 (1996): 476–500; John S. Dryzek, David Downs, Christian Hunold, and David Schlosberg, with Hans-Kristian Hernes, *Green States and Social Movements: Environmentalism in the United States, United Kingdom, Germany, and Norway* (New York: Oxford University Press, 2003); Dryzek, *Politics of the Earth*, chap. 8; Gonzalez, *Politics of Air Pollution*; George A. Gonzalez, *Urban Sprawl, Global Warming, and the Empire of Capital* (Albany: State University of New York, 2009); William J. Mitchell, Christopher E. Borroni-Bird, and Lawrence D. Burns, *Reinventing the Automobile: Personal Urban Mobility for the 21st Century* (Cambridge, MA: MIT Press, 2010).

62. World Business Council for Sustainable Development, *Pathways to 2050: Energy and Climate Change* (Washington, DC: World Business Council for Sustainable Development, 2007), 14–15.

63. Peter Newman and Jeffrey Kenworthy, *Sustainability and Cities: Overcoming Automobile Dependence* (Washington, DC: Island Press, 1999); Harriet Bulkeley and Michele M. Betsill, *Cities and Climate Change: Urban Sustainability and Global Environmental Governance* (New York: Routledge, 2003); Gonzalez, *Politics of Air Pollution*; and *Urban Sprawl, Global Warming, and the Empire of Capital*; Peter Newman, Timothy Beatley, and Heather Boyer, *Resilient Cities: Responding to Peak Oil and Climate Change* (Washington, DC: Island Press, 2009).

64. John S. Dryzek, *Rational Ecology: Environment and Political Economy* (New York: Blackwell, 1987).

65. Irvin C. Bupp and Jean-Claude Derian, *The Failed Promise of Nuclear Power: The Story of Light Water* (New York: Basic, 1978); Jon Gertner, "Atomic Balm?," *New York Times Magazine*, July 16, 2006, sec. 6, p. 36; Robert Vandenbosch and Susanne E. Vandenbosch, *Nuclear Waste Stalemate: Political and Scientific Controversies* (Salt Lake City: University of Utah Press, 2007); Max S. Power, *America's Nuclear Wastelands: Politics, Accountability, and Cleanup* (Pullman: Washington State University Press, 2008); Samuel Walker, *The Road to Yucca Mountain: The Development of Radioactive Waste Policy in the United States* (Los Angeles: University of California Press, 2009); Richard Burleson Stewart and Jane Bloom Stewart, *Fuel Cycle to Nowhere: U.S. Law and Policy on Nuclear Waste* (Nashville, TN: Vanderbilt University Press, 2011); Matthew L. Wald, "A Safer Nuclear Crypt," *New York Times*, July 6, 2011, B1.

66. William J. Broad, "For Iran, Enriching Uranium Only Gets Easier," *New York Times*, Mar. 8, 2010, D1; Matthew Kroenig, *Exporting the Bomb: Technology Transfer and the Spread of Nuclear Weapons* (Ithaca, NY: Cornell University Press, 2010).

67. One environmental liability of ethanol fuels is that more and more wilderness—especially in tropical rainforests—will be cleared to grow large amounts of source crops (e.g., sugar, soybeans, corn) to meet ethanol demand. Rick Barrett, "Ethanol Advocates Use Brazil as Model," *Milwaukee Journal Sentinel*, Mar. 6, 2007, D1; Stephen Leahy, "Biofuels Boom Spurring Deforestation," *Inter Press Service*, Mar. 21, 2007. (The destruction of wilderness to meet increased ethanol demand is predicted to elevate carbon dioxide levels in the atmosphere, as the burning/decomposition of cleared wilderness debris in and of itself emits large amounts of carbon dioxide. Also, wilderness—including savannah—is a key planetary carbon sink, and its destruction undermines the carbon cycle and the storage of carbon in a benign form.) Elisabeth Rosenthal, "Studies Call Biofuels a Greenhouse Threat," *New York Times*, Feb. 8, 2008, A9.

68. Stephen Dinan, "CBO Finds Greenhouse-gas Reduction Minuscule," *Washington Times*, Apr. 9, 2009, A7; Gerald Karey, "Ethanol Use Led to Higher Food Prices in US: CBO," *Platts Oilgram News*, Apr. 14, 2009, vol. 87, no. 72, Markets & Data sec., p. 10; Elisabeth Rosenthal, "Rush to Use Crops as Fuel Raises Food Prices and Hunger Fears," *New York Times*, Apr. 7, 2011, A1.

69. For a comprehensive discussion of the economic and environmental shortcomings of alternative fuels and carbon capture and sequestration technologies, see Paul Roberts, *The End of Oil: On the Edge of a Perilous New World* (New York: Houghton Mifflin, 2004); Richard Heinberg, *The Party's Over: Oil, War, and the Fate of Industrial Societies*, 2nd ed. (Gabriola Island, BC: New Society Publishers, 2005); Vaclav Smil, *Energy Myths and Realities: Bringing Science to the Energy Policy Debate* (Washington, DC: AEI Press,

2010); and *Energy Transitions: History, Requirements, Prospects* (Denver, CO: Praeger, 2010); Robin M. Mills, *Capturing Carbon: The New Weapon in the War Against Climate Change* (New York: Columbia University Press, 2011); Wolfgang Palz, ed., *Power for the World: The Emergence of Electricity from the Sun* (Singapore: Pan Stanford Publishing, 2011).

70. World Business Council for Sustainable Development, *Vision 2050: The New Agenda for Business* (Washington, DC: World Business Council for Sustainable Development, 2010), 28–29.

71. As quoted in Matthew L. Wald, "Will Hydrogen Clear the Air? Maybe Not, Say Some," *New York Times*, Nov. 12, 2003, C1.

72. Brian C. H. Steele and Angelika Heinzel, "Materials for Fuel-Cell Technologies," *Nature* 414 (Nov. 2001): 345; Joseph J. Romm, *The Hype about Hydrogen: Fact and Fiction in the Race to Save the Climate* (Washington, DC: Island Press, 2004).

73. World Business Council for Sustainable Development, *Pathways to 2050*, 8–9.

74. World Business Council for Sustainable Development, *Mobility 2030: Meeting the Challenge of Sustainability* (report overview) (Washington, DC: World Business Council for Sustainable Development, 2004), 25, emphasis added.

75. Ibid.

76. International Chamber of Commerce (ICC), *The International Chamber of Commerce* (Paris: International Chamber of Commerce, 2008), title page.

77. ICC (International Chamber of Commerce), What Is ICC?, http://www.iccwbo.org/id93/index.html (viewed Mar. 13, 2010).

78. ICC (International Chamber of Commerce), ICC Membership, http://www.iccwbo.org/id97/index.html (viewed Mar. 13, 2010).

79. ICC (International Chamber of Commerce), How ICC works, http://www.iccwbo.org/id96/index.html (viewed Mar. 13, 2010).

80. International Chamber of Commerce (ICC), Energy and the Environment: How Does it Work?, http://www.iccwbo.org/policy/environment/id1455/index.html (viewed Mar. 15, 2010).

81. International Chamber of Commerce (ICC), Commission on the Environment and Energy: Energy, http://www.iccwbo.org/policy/environment/id1461/index.html (viewed Mar. 15, 2010).

82. International Chamber of Commerce (ICC) Commission on Energy and Environment, *Energy Efficiency with Case Studies* (Paris: International Chamber of Commerce, 2009), 1.

83. International Chamber of Commerce (ICC), About BAE, http://www.iccwbo.org/BAE/id10994/index.html (viewed Mar. 15, 2010).

84. Keith Bradsher, *High and Mighty: SUVs—The World's Most Dangerous Vehicles and How They Got That Way* (New York: Public Affairs, 2002).

85. Jeffrey R. Kenworthy and Felix B. Laube, with Peter Newman, Paul Barter, Tamim Raad, Chamlong Poboon, and Benedicto Guia, Jr., *An International Sourcebook of Automobile Dependence in Cities 1960–1990* (Boulder: University Press of Colorado, 1999); Antonio Bento, Maureen L. Cropper, Ahmed Mushfiq Mobarak, and Katja Vinha, "The Effects of Urban Spatial Structure on Travel Demand in the United States," *Review of Economics and Statistics* 87, no. 3 (2005): 466–478; Jeffrey R. Kenworthy, "Sustainable Urban Transport: Developing Sustainability Rankings and Clusters Based on an International Comparison of Cities," in *Handbook of Sustainability Research*, vol. 20, Walter Leal Filho, ed. (New York: Peter Lang, 2005); Sabrina Tavernise and Robert Gebeloff, "Once Popular, Car Pools Go the Way of Hitchhiking," *New York Times*, Jan. 29, 2011, A1.

Table 6.1. Automobile Ownership and Use in Cities*

Region	Per Capita Automobile Ownership (per 1000 people)	Per Capita Automobile Use (kilometers)	Ratio of Per Capita U.S. Automobile Use Compared to Other Urban Areas
U.S. cities	587	18,155	—
Australian/NZ cities	575	11,387	1.60
Canadian cities	530	8,645	2.10
W. European cities	414	6,202	2.93
High Income Asian cities	210	3,614	5.02

Source: Jeffrey R. Kenworthy, "Energy Use and CO_2 Production in the Urban Passenger Transport Systems of 84 International Cities: Findings and Policy Implications," in *Urban Energy Transition From Fossil Fuels to Renewable Power*, ed. Peter Droege (Amsterdam: Elsevier, 2008), 211-36.

*Figures for most recent year available: 1995

US cities: Atlanta, Chicago, Denver, Houston, Los Angeles, New York, Phoenix, San Diego, San Francisco

Australian/New Zealand cities: Brisbane, Melbourne, Perth, Sydney, Wellington

Canadian cities: Calgary, Montreal, Ottawa, Toronto, Vancouver

Western Europe cities: Graz, Vienna, Brussels, Copenhagen, Helsinki, Lyon, Nantes, Paris, Marseilles, Berlin, Frankfurt, Hamburg, Dusseldorf, Munich, Ruhr, Stuttgart, Athens, Milan, Bologna, Rome, Amsterdam, Oslo, Barcelona, Madrid, Stockholm, Bern, Geneva, Zurich, London, Manchester, Newcastle, Glasglow

High Income Asia cities: Osaka, Sapporo, Tokyo, Hong Kong, Singapore, Taipei

Conclusion

1. David Lochbaum, *The NRC and Nuclear Power Plant Safety in 2010: A Brighter Spotlight Needed* (Cambridge, MA: Union of Concerned Scientists, 2011); Matthew L. Wald, "Japan Orders Evacuation Near 2nd Nuclear Plant," *New York Times*, Mar. 12, 2011, A10; Henry Fountain, "A Look at the Mechanics of a Partial Meltdown," *New York Times*, Mar. 14, 2011, A9; "Analysis: Japan Nuclear Crisis Passes Three Mile," Reuters, Mar. 15, 2011; Tom Zeller, Jr., "With U.S. Nuclear Plants under Scrutiny, Too, a Report Raises Safety Concerns," *New York Times*, Mar. 18, 2011, A11; Hiroko Tabuchi, Norimitsu Onishi, and Ken Belson, "Japan Extended Reactor's Life, Despite Warning," *New York Times*, Mar. 22, 2011, A1; William J. Broad, "From Far Labs, a Vivid Picture Emerges of Japan Crisis," *New York Times*, Apr. 3, 2011, A1; Hiroko Tabuchi, "Japanese Workers Braved Radiation for a Temp Job," *New York Times*, Apr. 10, 2011, A1; Keith Bradsher, Hiroko Tabuchi, and Andrew Pollack, "Japanese Officials on Defensive as Nuclear Alert Level Rises," *New York Times*, Apr. 13, 2011, A5; Hiroko Tabuchi, "Fears of Fission Rise at Stricken Japanese Plant," *New York Times*, Nov. 3, 2011, A17.

2. John M. Broder, "U.S. Nuclear Industry Faces New Uncertainty," *New York Times*, Mar. 14, 2011, A1; Judy Dempsey and Sharon LaFraniere, "In Europe and China, Japan's Crisis Renews Fears about Nuclear Power," *New York Times*, Mar. 17, 2011, B4.

The *Economic Intelligence Unit*, of the periodical *The Economist*, predicts that in spite of the political fallout of the Japan Fukushima nuclear disaster, the world's nuclear power capacity will continue to grow. *Economic Intelligence Unit, The Future of Nuclear Energy One Step Back, Two Steps Forward*, June 2011, www.eiu.com. According to the World Nuclear Association, an industry group, nuclear reactors provide 15 percent of global energy, and sixty more nuclear plants are presently being built. Judy Dempsey, "Panel Urges Germany to Close Nuclear Plants by 2021," *New York Times*, May 12, 2011, B7.

3. Hiroko Tabuchi, "Japan Premier Wants Shift Away from Nuclear Power," *New York Times*, July 14, 2011, A6; Elisabeth Rosenthal, "Germany Dims Nuclear Plants, but Hopes to Keep Lights On," *New York Times*, Aug. 30, 2011, A1.

4. Jeffrey R. Kenworthy, "Energy Use and CO2 Production in the Urban Passenger Transport Systems of 84 International Cities: Findings and Policy Implications," in *Urban Energy Transition from Fossil Fuels to Renewable Power*, Peter Droege, ed. (Amsterdam: Elsevier, 2008).

5. John M. Broder, "The Year of Peril and Promise in Energy Production," *International Herald Tribune*, Oct. 11, 2011, Finance sec., 19; David Toke, *Ecological Modernisation and Renewable Energy* (New York: Palgrave Macmillan, 2011).

6. Stephen Kurczy, "International Energy Agency Says 'Peak Oil' Has Hit," *Christian Science Monitor*, Nov. 11, 2010.

7. David Jolly, "Booming Car Sales in China May Bypass Diesels," *New York Times*, Oct. 1, 2010.

8. Robert F. Worth, "Unrest Encircles Saudis, Stoking Sense of Unease," *New York Times*, Feb. 20, 2011, A13; Clifford Krauss and Christine Hauser, "Oil Soars as Libyan Furor Shakes Markets," *New York Times*, Feb. 23, 2011, A1; Clifford Krauss, "Why the Disruption of Libyan Oil Has Led to a Price Spike," *New York Times*, Feb. 24, 2011, B1; Steven Lee Myers, "Tumult of Arab Spring Prompts Worries in Washington," *New York Times*, Sept. 18, 2011, A1.

9. Jad Mauawad and Clifford Krauss, "Tremors from Libya Threaten to Rattle the Oil World," *New York Times*, Feb. 28, 2011, B1.

10. There is a scientific consensus that the average global temperature should not increase more than 2°C above the preindustrial average. Participants of the 2009 Copenhagen Climate Change Conference agreed to this goal. A serious problem is that it is unlikely we can keep global temperature increases down to 2°C, as we have already seen temperature increase by 0.76°C, and even if we maintain atmospheric carbon dioxide levels at the 2000 levels, that would nevertheless add at least another 0.6°C. Thus, we are already up to nearly 1.4°C in global temperature increase. Catherine Gautier, *Oil, Water, and Climate: An Introduction* (New York: Cambridge University Press, 2008); Catherine Gautier and Jean-Louis Fellous, eds., *Facing Climate Change Together* (New York: Cambridge University Press, 2008); John Houghton, *Global Warming: The Complete Briefing*, 4th ed. (New York: Cambridge University Press, 2009); Mark Maslin, *Global Warming: A Very Short Introduction* (New York: Oxford University Press, 2009).

11. Andrés Cala, "Drive Toward Low-Carbon Future Stalls," *New York Times*, Oct. 25, 2011. Web; John M. Broder, "At Meeting on Climate Change, Urgent Issues but Low Expectations," *New York Times*, Nov. 28, 2011, A8; Justin Gillis, "Global Carbon Dioxide Emissions in 2010 Show the Biggest Jump Ever Recorded," *New York Times*, Dec. 5, 2011, A4; John M. Broder, "Climate Talks in Durban Yield Limited Agreement," *New York Times*, Dec. 12, 2011, A9. The *New York Times* reported, in 2012, that "if nothing changes, clean energy funding [from the U.S. federal government] will drop from a peak of $44.3 billion in 2009 to $16 billion this year and $11 billion in 2014—a 75 percent decline." "The End of Clean Energy Subsidies?" *New York Times*, May 6, 2012, SR12.

Bibliography

Abramson, Rudy, *Spanning the Century: The Life of W. Averell Harriman, 1891–1986* (New York: W. Morrow, 1992).

Allardice, Corbin, and Edward R. Trapnell, *The Atomic Energy Commission* (New York: Praeger, 1974).

Allin, Dana, and Steven Simon, *The Sixth Crisis: Iran, Israel, America, and the Rumors of War* (New York: Oxford University Press, 2010).

American Petroleum Institute, *Petroleum Facts and Figures: Centennial Edition* (New York: American Petroleum Institute, 1959).

American Road Congress, *Papers, Addresses, and Resolutions before the American Road Congress, Richmond, Virginia, November 20–23, 1911* (Baltimore: Waverly Press, 1911).

"Analysis: Japan Nuclear Crisis Passes Three Mile," Reuters, Mar. 15, 2011.

Appelbaum, Binyamin, "Without Loan Giants, 30-Year Mortgage May Fade Away," *New York Times*, Mar. 4, 2011, A1.

Armand, Louis, *Some Aspects of the European Energy Problem: Suggestions for Collective Action* (Paris: Organization for European Cooperation, 1955).

Aronowitz, Stanley, and Peter Bratsis, eds., *Paradigm Lost: State Theory Reconsidered* (Minneapolis: University of Minnesota Press, 2002).

"As Oil Consultant, He's Without Like or Equal," *New York Times*, July 27, 1969, sec. 3, p. 3.

Atkinson, Robert D., *The Past and Future of America's Economy: Long Waves of Innovation that Power Cycles of Growth* (Northampton, MA: Edward Elgar, 2004).

Austen, Ian, "Uranium Processor Still Optimistic about Nuclear Industry," *New York Times*, Mar. 26, 2011, B3.

Bajaj, Vikas, "Resistance to Jaitapur Nuclear Plant Grows in India," *New York Times*, Apr. 14, 2011, B1.

———, "India's Investment in the Sun," *New York Times*, Dec. 29, 2011, B1.

Baker, Peter, "Arms Talks Now Turn to Short-Range Weapons," *New York Times*, Dec. 25, 2010, A4.

133

Balint, Benjamin, *Running Commentary: The Contentious Magazine That Transformed the Jewish Left Into the Neoconservative Right* (New York: PublicAffairs, 2010).

Balogh, Brian, *Chain Reaction: Expert Debate and Public Participation in American Commercial Nuclear Power, 1945-1975* (New York: Cambridge University Press, 1991).

Banks, Ferdinand E., *The Political Economy of Oil* (Lexington, MA: Lexington Books, 1980).

Barach, Arnold B., and the Twentieth Century Fund, *USA and Its Economic Future: A Twentieth Century Fund Survey* (New York: Macmillan, 1964).

Bardou, Jean-Pierre, Jean-Jacques Chanaron, Patrick Fridenson, and James M. Laux, *The Automobile Revolution: The Impact of an Industry* (Chapel Hill: University of North Carolina Press, 1982).

Barker, T. C., "The International History of Motor Transport," *Journal of Contemporary History* 20, no. 1 (1985): 3–19.

Barrett, Paul, *The Automobile and Urban Transit* (Philadelphia: Temple University Press, 1983).

Barrett, Rick, "Ethanol Advocates Use Brazil as Model," *Milwaukee Journal Sentinel*, Mar. 6, 2007, D1.

Barringer, Felicity, "A Coalition for Firm Limit on Emissions," *New York Times*, Jan. 19, 2007, C1.

———, "With Push Toward Renewable Energy, California Sets Pace for Solar Power," *New York Times*, July 16, 2009, A19.

Barrow, Clyde W., *Universities and the Capitalist State: Corporate Liberalism and the Reconstruction of American Higher Education, 1894–1928* (Madison: University of Wisconsin Press, 1990).

———, "Corporate Liberalism, Finance Hegemony, and Central State Intervention in the Reconstruction of American Higher Education," *Studies in American Political Development* 6 (fall 1992): 420–444.

———, *Critical Theories of the State* (Madison: Wisconsin University Press, 1993).

Barry, Ellen, "Lessons from Chernobyl for Japan," *New York Times*, Mar. 20, 2011, WK1.

Baumgartner, Frank R., Jeffrey M. Berry, Marie Hojnacki, David C. Kimball, and Beth L. Leech, *Lobbying and Policy Change: Who Wins, Who Loses, and Why* (Chicago: University of Chicago Press, 2009).

Baxandall, Rosalyn, and Elizabeth Ewen, *Picture Windows: How the Suburbs Happened* (New York: Basic Books, 2000).

Beaubouef, Bruce A., *The Strategic Petroleum Reserve: U.S. Energy Security and Oil Politics, 1975–2005* (College Station: Texas A&M University Press, 2007).

Beauregard, Robert A., *When America Became Suburban* (Minneapolis: University of Minnesota Press, 2006).

Becker, Helmut, *High Noon in the Automotive Industry* (New York: Springer, 2006).

Becker, Jo, "U.S. Approved Business With Blacklisted Nations," *New York Times*, Dec. 24, 2010, A1.

Becker, Jo, and William J. Broad, "New Doubts about Turning Plutonium Into a Fuel," *New York Times*, Apr. 11, 2011, A14.

Becker, William H., and William M. McClenahan, Jr., *The Market, the State, and the Export-Import Bank of the United States, 1934-2000* (Cambridge: Cambridge University Press, 2003).

Belford, Aubrey, "Indonesia to Continue Plans for Nuclear Power," *New York Times*, Mar. 18, 2011, B5.

Bellany, Ian, Coit D. Blacker, and Joseph Gallacher, *The Nuclear Non-proliferation Treaty* (New York: Routledge, 1985).

Belson, Ken, Keith Bradsher, and Matthew L. Wald, "Executives May Have Lost Valuable Time at Damaged Nuclear Plant," *New York Times*, Mar. 20, 2011, A12.

Bento, Antonio, Maureen L. Cropper, Ahmed Mushfiq Mobarak, and Katja Vinha, "The Effects of Urban Spatial Structure on Travel Demand in the United States," *The Review of Economics and Statistics* 87, no. 3 (2005): 466–478.

Berman, Edward H., *Influence of the Carnegie, Ford, and Rockefeller Foundations on American Foreign Policy* (Albany: State University of New York Press, 1983).

Bernstein, Michael A., *The Great Depression: Delayed Recovery and Economic Change in America, 1929–1939* (New York: Cambridge University Press, 1987).

Bill, James A., *The Eagle and the Lion: The Tragedy of American-Iranian Relations* (New Haven, CT: Yale University Press, 1988).

Blair, John M., *The Control of Oil* (New York: Pantheon, 1976).

Blas, Javier, James Boxell, and Kevin Morrison, "Energy Investment Too Small to Meet Growth in Demand, Warns Watchdog," *Financial Times*, May 4, 2005, 1.

Blatt, Harvey, *America's Environmental Report Card: Are We Making the Grade?* (Cambridge, MA: MIT Press, 2005).

———, *America's Environmental Report Card: Are We Making the Grade*, 2nd ed. (Cambridge, MA: MIT Press, 2011).

Block, Fred, "Understanding the Diverging Trajectories of the United States and Western Europe: A Neo-Polanyian Analysis," *Politics & Society* 35, no. 1 (2007): 3–33.

Bloom, Nicholas, *Public Housing that Worked: New York in the Twentieth Century* (Philadelphia: University of Pennsylvania Press, 2008).

Bogart, William T., *Don't Call It Sprawl: Metropolitan Structure in the Twenty-First Century* (New York: Cambridge University Press, 2006).

Borden, William S., *The Pacific Alliance: United States Foreign Economic Policy and Japanese Trade Recovery, 1947–1955* (Madison: University of Wisconsin Press, 1984).

Bordo, Michael D., Claudia Goldin, and Eugene N. White, eds., *The Defining Moment: The Great Depression and the American Economy in the Twentieth Century* (Chicago: University of Chicago Press, 1998).

Bottles, Scott, *Los Angeles and the Automobile: The Making of the Modern City* (Los Angeles: University of California Press, 1987).

Boudette, Neal E., and Norihiko Shirouzu, "Car Makers' Boom Years Now Look Like a Bubble," *Wall Street Journal*, May 20, 2008, A1.

Bradford, Travis, *Solar Revolution: The Economic Transformation of the Global Energy Industry* (Cambridge, MA: MIT Press, 2006).

Bradsher, Keith, *High and Mighty: SUVs—The World's Most Dangerous Vehicles and How They Got That Way* (New York: Public Affairs, 2002).

———, "China Racing Ahead of U.S. in the Drive to Go Solar," *New York Times*, Aug. 25, 2009, A1.

———, "Nuclear Power Expansion in China Stirs Concerns," *New York Times*, Dec. 16, 2009, A1.

———, "China Leading Global Race to Make Clean Energy," *New York Times*, Jan. 30, 2010, A1

———, "On Clean Energy, China Skirts Rules," *New York Times*, Sept. 9, 2010, A1.

———, "Solar Panel Maker Moves Work to China," *New York Times*, Jan. 15, 2011, B1.

———, "A Radical Kind of Reactor," *New York Times*, Mar. 25, 2011, B1.

———, "U.S. Posted a Trade Surplus in Solar Technologies, Study Finds," *New York Times*, Aug. 29, 2011, B4.

———, "China Benefits as U.S. Solar Industry Withers," *New York Times*, Sept. 2, 2011, B1.

———, "200 Chinese Subsidies Violate Rules, U.S. Says," *New York Times*, Oct. 7, 2011, B3.

———, "U.S. Solar Firms Accuse Chinese of Trade Violations," *New York Times*, Oct. 20, 2011, B1.

———, "China Bends to U.S. Complaint on Solar Panels but Plans Retaliation," *New York Times*, Nov. 22, 2011, B7.

Brenner, Michael J., *Nuclear Power and Non-Proliferation: The Remaking of U.S. Policy* (New York: Cambridge University Press, 1981).

Brenner, Robert, *The Boom and the Bubble: The U.S. in the World Economy* (New York: Verso, 2002).

————, "New Boom or New Bubble: The Trajectory of the U.S. Economy," *New Left Review* 25 (Jan./Feb. 2004): 57–102.

Broad, William J., "For Iran, Enriching Uranium Only Gets Easier," Mar. 8, 2010, *New York Times*, D1.

————, "Buffett Helps Create Nuclear Fuel Bank," *New York Times*, Dec. 4, 2010, A4.

————, "From Far Labs, a Vivid Picture Emerges of Japan Crisis," *New York Times*, Apr. 3, 2011, A1.

Broad, William J., and David E. Sanger, "C.I.A. Secrets Could Surface in Swiss Nuclear Case," *New York Times*, Dec. 24, 2010, A1.

Broder, John M., "House, 314–100, Passes Broad Energy Bill," *New York Times*, Dec. 19, 2007, A24.

————, "Climate Talks End with Modest Deal on Emissions," *New York Times*, Dec. 12, 2010, A16.

————, "Director of Policy on Climate Will Leave, Her Goal Unmet," *New York Times*, Jan. 25, 2011, A15.

————, "House Panel Votes to Strip E.P.A. of Power to Regulate Greenhouse Gases," *New York Times*, Mar. 11, 2011, A18.

————, "U.S. Nuclear Industry Faces New Uncertainty," *New York Times*, Mar. 14, 2011, A1.

————, "The Year of Peril and Promise in Energy Production," *International Herald Tribune*, Oct. 11, 2011, Finance sec., 19.

————, "At Meeting on Climate Change, Urgent Issues but Low Expectations," *New York Times*, Nov. 28, 2011, A8.

————, "Climate Talks in Durban Yield Limited Agreement," *New York Times*, Dec. 12, 2011, A9.

Broder, John M., and Elisabeth Rosenthal, "Obama Has Goal to Wrest a Deal in Climate Talks," *New York Times*, Dec. 18, 2009, A1.

Broder, John M., and Matthew L. Wald, "Report Blasts Management Style of Nuclear Regulatory Commission Chairman," *New York Times*, June 11, 2011, A13.

Bromley, Simon, *American Hegemony and World Oil: The Industry, the State System and the World Economy* (University Park: Pennsylvania State University Press, 1991).

Bronson, Rachel, *Thicker than Oil: America's Uneasy Partnership with Saudi Arabia* (New York: Oxford University Press, 2006).

Brown, Donald A., *American Heat: Ethical Problems with the United States' Response to Global Warming* (Lanham, MD: Rowman and Littlefield, 2002).

Brown, Marilyn A. and Benjamin K. Sovacool, *Climate Change and Global Energy Security: Technology and Policy Options* (Cambridge, MA: MIT Press, 2011).

Buenger, Walter L., and Joseph A. Pratt, *But Also Good Business: Texas Commerce Banks and the Financing of Houston and Texas, 1886–1986* (College Station: Texas A&M University Press, 1986).

Bulkeley, Harriet, and Michele M. Betsill, *Cities and Climate Change: Urban Sustainability and Global Environmental Governance* (New York: Routledge, 2003).

Bunkley, Nick, "Toyota Ahead of G.M. in 2008 Sales," *New York Times*, Jan. 22, 2009, B2.

Bupp, Irvin C., and Jean-Claude Derian, *The Failed Promise of Nuclear Power: The Story of Light Water* (New York: Basic Books, 1978).

Burn, Duncan Lyall, *Nuclear Power and the Energy Crisis: Politics and the Atomic Industry* (New York: New York University Press, 1978).

Büthe, Tim, "Taking Temporality Seriously: Modeling History and the Use of Narratives as Evidence," *American Political Science Review* 96, no. 3 (2002): 481–493.

Cala, Andrés, "Iraq Struggles with High-Flying Oil Goals," *New York Times*, Feb. 22, 2011. Web.

———, "Drive Toward Low-Carbon Future Stalls," *New York Times*, Oct. 25, 2011. Web.

Caldicott, Helen, *Nuclear Power Is Not the Answer* (New York: New Press, 2006).

Calmes, Jackie, "Leader Picked for Review of U.S. Loans on Energy," *New York Times*, Oct. 29, 2011, A15.

Calthorpe, Peter, *Urbanism in the Age of Climate Change* (Washington, DC: Island Press, 2011).

Camilleri, Joseph A., *The State and Nuclear Power: Conflict and Control in the Western World* (Seattle: University of Washington Press, 1984).

Campbell, John L., *Nuclear Power and the Contradiction of U.S. Policy* (Ithaca, NY: Cornell University Press, 1988).

Carpenter, Daniel P., *The Forging of Bureaucratic Autonomy: Reputations, Networks, and Policy Innovations in Executive Agencies, 1862–1928* (Princeton, NJ: Princeton University Press, 2001).

Carskadon, Thomas Reynolds, and George Henry Soule, *USA in New Dimensions: The Measure and Promise of America's Resources, A Twentieth Century Fund Survey* (New York: Macmillan, 1957).

Case, Josephine Young, *Owen D. Young and American Enterprise: A Biography* (Boston: David R. Godine, 1982).

Cass, Loren R., *The Failures of American and European Climate Policy: International Norms, Domestic Politics, and Unachievable Commitments* (Albany: State University of New York Press, 2006).

Carter, Luther J., *Nuclear Imperatives and Public Trust: Dealing with Radioactive Waste* (Washington, DC: Resources for the Future, 1987).

Chan, Sewell, "U.S. Says China Fund Breaks Rules," *New York Times*, Dec. 23, 2010, B1.

Chernus, Ira, *Eisenhower's Atoms for Peace* (College Station: Texas A&M University Press, 2002).

Christoff, Peter, "Ecological Modernization, Ecological Modernities," *Environmental Politics* 5, no. 3 (1996): 476–500.

Cohen, Lizabeth, "Is There an Urban History of Consumption?," *Journal of Urban History* 29, no. 2 (2003): 87–106.

Cohen, Michael P., *The History of the Sierra Club, 1892–1970* (San Francisco: Sierra Club Books, 1988).

Collier, Peter, and David Horowitz, *The Rockefellers: An American Dynasty* (New York: Holt, Rinehart, and Winston, 1976).

Collins, Catherine, and Douglas Frantz, *Fallout: The True Story of the CIA's Secret War on Nuclear Trafficking* (New York: Free Press, 2011).

Commission for Energy, *Europe's Growing Needs of Energy: How Can They Be Met?* (Paris: Organisation for European Economic Co-Operation, 1956).

Compston, Hugh, and Ian Bailey, eds., *Turning Down the Heat: The Politics of Climate Policy in Affluent Democracies* (New York: Palgrave Macmillan, 2008).

Conybeare, John A. C., *Merging Traffic: The Consolidation of the International Automobile Industry* (Lanham, MD: Rowman & Littlefield, 2004).

Cooper, Michael, and Dalia Sussman, "Nuclear Power Loses Support in New Poll," *New York Times*, Mar. 23, 2011, A15.

Cyphers, Christopher, J., *The National Civic Federation and the Making of New Liberalism, 1900–1915* (Westport, CT: Praeger, 2002).

Dahl, Robert A., and Charles E. Lindblom, *Politics, Economics, and Welfare* (New Haven, CT: Yale University Press, 1953).

———, preface to *Politics, Economics, and Welfare* (New Haven, CT: Yale University Press, 1976).

Dahl, Robert A., *A Preface to Democratic Theory* (Chicago: University of Chicago Press, 1956).

———, *Who Governs? Democracy and Power in an American City* (New Haven, CT: Yale University Press, 1961 [2005]).

Daily, Matt, and Sarah McBride, "Financing Dearth Holds Solar Back in U.S.," *New York Times*, Oct. 17, 2010. Web.

Davis, David, *Energy Politics* (New York: St. Martin's Press, 1993).

Davis, Stacy C., Susan W. Diegel, and Robert G. Boundy, *Transportation Energy Data Book*, 29th ed. (Washington, DC: Department of Energy, 2010).

———, *Transportation Energy Data Book*, 30th ed. (Washington, DC: Department of Energy, 2011).

Dawson, Frank G., *Nuclear Power: Development and Management of a Technology* (Seattle: University of Washington Press, 1976).

Dawson, Jane, and Robert Darst, "Meeting the Challenge of Permanent Nuclear Waste Disposal in an Expanding Europe: Transparency, Trust and Democracy," *Environmental Politics* 15, no. 4 (2006): 610–627.

de Perthuis, Christian, *Economic Choices in a Warming World* (New York: Cambridge University Press, 2011).

Dean, Cornelia, "Group Urges Research into Aggressive Efforts to Fight Climate Change," *New York Times*, Oct. 4, 2011, A18.

Deffeyes, Kenneth S., *Hubbert's Peak: The Impending World Oil Shortage* (Princeton, NJ: Princeton University Press, 2001).

———, *Beyond Oil: The View from Hubbert's Peak* (New York: Hill and Wang, 2005).

———, *When Oil Peaked* (New York: Hill and Wang, 2010).

Dempsey, Judy, "Germany Extends Nuclear Plants' Life," *New York Times*, Sept. 6, 2010. Web.

———, "Panel Urges Germany to Close Nuclear Plants by 2021," *New York Times*, May 12, 2011, B7.

Dempsey, Judy and Sharon LaFraniere, "In Europe and China, Japan's Crisis Renews Fears about Nuclear Power," *New York Times*, Mar. 17, 2011, B4.

Deutch, John M., *The Crisis in Energy Policy* (Cambridge, MA: Harvard University Press, 2011).

Dewan, Shaila, and Robert Gebeloff, "One Percent, Many Variations," *New York Times*, Jan. 15, 2012, A1.

Dewhurst, Frederic, and the Twentieth Century Fund, *America's Needs and Resources: A New Survey* (New York: Twentieth Century Fund, 1955).

Domhoff, G. William, *The Bohemian Grove and Other Retreats* (New York: Harper and Row, 1974).

———, *The Powers That Be* (New York: Random House, 1978).

———, *Who Really Rules: New Haven and Community Power Reexamined* (Santa Monica, CA: Goodyear, 1978).

———, *The Power Elite and the State* (New York: Aldine de Gruyter, 1990).

———, *State Autonomy or Class Dominance?* (New York: Aldine de Gruyter, 1996).

———, *Who Rules America?*, 6th ed. (New York: McGraw-Hill, 2009).

Domhoff, G. William and Michael J. Webber, *Class and Power in the New Deal* (Palo Alto, CA: Stanford University Press, 2011).

Dorrien, Gary, *The Neoconservative Mind: Politics, Culture, and the War of Ideology*, (Philadelphia: Temple University Press, 1993).

———, *Imperial Designs: Neoconservatism and the New Pax Americana* (New York: Routledge, 2004).

Dowie, Mark, *Losing Ground: American Environmentalism at the Close of the Twentieth Century* (Cambridge, MA: MIT Press, 1995).

———, *American Foundations: An Investigative History* (Cambridge, MA: MIT Press, 2001).

Drolet, Jean-François, *American Neoconservatism: The Politics and Culture of a Reactionary Idealism* (New York: Columbia University Press, 2011).

Dryzek, John S., *Rational Ecology: Environment and Political Economy* (New York: Blackwell, 1987).

———, *Democracy in Capitalist Times* (New York: Oxford University Press, 1996).

———, *The Politics of the Earth*, 2nd ed. (New York: Oxford University Press, 2005).

Dryzek, John S., David Downs, Christian Hunold, and David Schlosberg, with Hans-Kristian Hernes, *Green States and Social Movements: Environmentalism in the United States, United Kingdom, Germany, and Norway* (New York: Oxford University Press, 2003).

Du Boff, Richard B., *Accumulation and Power: An Economic History of the United States* (Armonk, NY: M.E. Sharpe, 1989).

Duffield, John S., *Over a Barrel: The Costs of U.S. Foreign Oil Dependence* (Stanford, CA: Stanford University Press, 2008).

Duffy, Robert J., *Nuclear Politics in America: A History and Theory of Government Regulation* (Lawrence: University of Kansas Press, 1997).

Dunn, James, *Miles to Go: European and American Transportation Policies* (Cambridge, MA: MIT Press, 1981).

———, *Driving Forces: The Automobile, Its Enemies, and the Politics of Mobility* (Washington, DC: Brookings Institution Press, 1998).

Eakins, David, "Business Planners and America's Postwar Expansion," in *Corporations and the Cold War*, David Horowitz, ed. (New York: Monthly Review Press, 1969).

———, "Policy-Planning for the Establishment," in *A New History of Leviathan*, Ronald Radosh and Murray N. Rothbard, eds. (New York: E.P. Dutton & Co., 1972).

Eckstein, Rick, *Nuclear Power and Social Power* (Philadelphia: Temple University Press, 1997).

Economic Intelligence Unit, The Future of Nuclear Energy One Step Back, Two Steps Forward, June 2011, www.eiu.com.

Ehrman, John, *The Rise of Neoconservatism: Intellectuals and Foreign Affairs*, 1945–1994 (New York: Cambridge University Press, 1995).

Elliot, David, ed., *Nuclear or Not? Does Nuclear Power Have a Place in a Sustainable Energy Future?* (New York: Palgrave MacMillan, 2007).

Energy Advisory Commission, *Towards a New Energy Pattern in Europe* (Paris: Organisation for European Economic Co-operation, 1960).

Energy Information Administration, *Annual Energy Review 2010* (Washington, DC: U.S. Department of Energy, 2011).

Erlanger, Steven, "French Plans For Energy Reaffirm Nuclear Path," *New York Times*, Aug. 17, 2008, A6.

———, "With Prospect of U.S. Slowdown, Europe Fears a Worsening Debt Crisis," *New York Times*, Aug. 8, 2011, B3.

Fackler, Martin, "Toyota's Profit Soars, Helped By U.S. Sales," *New York Times*, Aug. 5, 2006, C4.

———, "Toyota Expects Decline in Annual Profit," *New York Times*, May 9, 2008, C3.

Falk, Jim, *Global Fission: The Battle over Nuclear Power* (New York: Oxford University Press, 1982).

Fearon, Peter, *War, Prosperity, and Depression: The U.S. Economy 1917–45* (Lawrence: University Press of Kansas, 1987).

Ferguson, Charles D., *Nuclear Energy: What Everyone Needs to Know* (New York: Oxford University Press, 2011).

Field, Alexander J., "Technological Change and U.S. Productivity Growth in the Interwar Years," *Journal of Economic History* 66, no. 1 (2006): 203–234.

Finegold, Kenneth, and Theda Skocpol, *State and Party in America's New Deal* (Madison: University of Wisconsin Press, 1995).

Findlay, Trevor, *Nuclear Energy and Global Governance* (New York: Routledge, 2011).

Fisher, Dana, *National Governance and the Global Climate Change Regime* (Lanham, MD: Rowman and Littlefield, 2004).

Fishman, Robert, *Bourgeois Utopias: The Rise and Fall of Suburbia* (New York: Basic Books, 1987).

Flink, James, *The Car Culture* (Cambridge, MA: MIT Press, 1975).

———, *The Automobile Age* (Cambridge, MA: MIT Press, 1990).

Fogelson, Robert M., *Bourgeois Nightmares: Suburbia, 1870–1930* (New Haven, CT: Yale University Press, 2005).

Ford, Daniel, *The Cult of the Atom: The Secret Papers of the Atomic Energy Commission* (New York: Simon and Schuster, 1982).

Fosdick, Raymond B., *The Story of the Rockefeller Foundation* (New York: Transaction Publishers, 1988 [1952]).

Foss, Michelle Michot, "Ignorance Stifles Innovation in Solving Energy Problems," *International Herald Tribune*, Oct. 12, 2011, Finance sec., 11.

Fountain, Henry, "A Look at the Mechanics of a Partial Meltdown," *New York Times*, Mar. 14, 2011, A9.

Freese, Barbara, *Coal: A Human History*, 4th ed. (New York: Penguin, 2004).

French, Michael, *U.S. Economic History since 1945* (Manchester: Manchester University Press, 1997).

Friedman, Murray, *The Neoconservative Revolution: Jewish Intellectuals and the Shaping of Public Policy* (New York: Cambridge University Press, 2005).

Frumkin, Norman, *Tracking America's Economy*, 4th ed. (Armonk, NY: M.E. Sharpe, 2004).

Funigiello, Philip J., *American-Soviet Trade in the Cold War* (Chapel Hill: University of North Carolina Press, 1988).

Galbraith, Kate, "Natural Gas, Scrutinized, Pushes for Growth," *New York Times*, Mar. 11, 2011, A21.

————, "Certainties of 1970s Energy Crisis Have Fallen Away," *New York Times*, Apr. 3, 2011, A25.

————, "Wind Power Gains as Gear Improves," *New York Times*, Aug. 7, 2011. Web.

————, "Future of Solar and Wind Power May Hinge on Federal Aid, *New York Times*, Oct. 26, 2011, F5.

————, "A New Urgency to the Problem of Storing Nuclear Waste," *International Herald Tribune*, Nov. 28, 2011, Finance sec., 20.

Gautier, Catherine, *Oil, Water, and Climate: An Introduction* (New York: Cambridge University Press, 2008).

Gautier, Catherine, and Jean-Louis Fellous, eds., *Facing Climate Change Together* (New York: Cambridge University Press, 2008).

General Accounting Office, *Energy Task Force: Process Used to Develop the National Energy Policy* (Washington, DC: Government Printing Office, 2003).

Gertner, Jon, "Atomic Balm?," *New York Times Magazine*, July 16, 2006, sec. 6, p. 36.

Gillis, Justin, "Global Carbon Dioxide Emissions in 2010 Show the Biggest Jump Ever Recorded," *New York Times*, Dec. 5, 2011, A4.

Gonzalez, George A., *Corporate Power and the Environment: The Political Economy of U.S. Environmental Policy* (Lanham, MD: Rowman & Littlefield, 2001).

————, "Ideas and State Capacity, or Business Dominance? A Historical Analysis of Grazing on the Public Grasslands," *Studies in American Political Development* 15 (fall 2001): 234–244.

————, "The Comprehensive Everglades Restoration Plan: Economic or Environmental Sustainability?," *Polity* 37, no. 4 (2005): 466–490.

————, *The Politics of Air Pollution: Urban Growth, Ecological Modernization, and Symbolic Inclusion* (Albany: State University of New York Press, 2005).

————, "Urban Sprawl, Global Warming, and the Limits of Ecological Modernization," *Environmental Politics* 14, no. 3 (2005): 344–362.

————, "An Eco-Marxist Analysis of Oil Depletion via Urban Sprawl," *Environmental Politics* 15, no. 4 (2006): 515–531.

————, "The Wilderness Act of 1964 and the Wilderness Preservation Policy Network," *Capitalism Nature Socialism* 20, no. 4 (2009): 31–52.

————, *Urban Sprawl, Global Warming, and the Empire of Capital* (Albany: State University of New York Press, 2009).

Goodell, Jeff, *Big Coal: The Dirty Secret behind America's Energy Future* (New York: Mariner, 2007).

Goodman, Peter S., "The Economy: Trying to Guess What Happens Next." *New York Times*, Nov. 25, 2007, sec. 4, p. 1.

Goodstein, David, *Out of Gas: The End of the Age of Oil* (New York: Norton, 2004).

Gordon, Colin, *New Deals: Business, Labor, and Politics in America, 1920–1935* (New York: Cambridge University Press, 1994).

Graetz, Michael J., *The End of Energy: The Unmaking of America's Environment, Security, and Independence* (Cambridge, MA: MIT Press, 2011).

Gregor, Allison, "Idle Land Finds a Purpose as Farms for Solar Power," *New York Times*, Mar. 23, 2011, B7.

Guilhot, Nicolas, ed., *The Invention of International Relations Theory: Realism, the Rockefeller Foundation, and the 1954 Conference on Theory* (New York: Columbia University Press, 2011).

Hajer, Maarten A., *The Politics of Environmental Discourse* (New York: Oxford University Press, 1995).

Halper, Stefan, and Jonathan Clarke, *America Alone: The Neo-Conservatives and the Global Order* (Cambridge: Cambridge University Press, 2004).

Hamdan, Sara, "Energy Plant Makes a Leap in Solar Power," *New York Times*, Oct. 25, 2011. Web.

Harrington, Winston, and Virginia McConnell, *Resources for the Future Report: Motor Vehicles and the Environment* (Washington, DC: Resources for the Future, 2003).

Hatch, Michael T., *Politics and Nuclear Power: Energy Policy in Western Europe* (Lexington: University Press of Kentucky, 1986).

———, ed., *Environmental Policymaking: Assessing the Use of Alternative Policy Instruments* (Albany: State University of New York Press, 2005).

Haugland, Torleif, Helge Ole Bergensen, and Kjell Roland, *Energy Structures and Environmental Futures* (New York: Oxford University Press, 1998).

Hay, Colin, Michael Lister, and David Marsh, eds., *The State: Theories and Issues* (New York: Palgrave Macmillan, 2006).

Hays, Samuel, "The Politics of Reform in Municipal Government in the Progressive Era," *Pacific Northwest Quarterly* 55, no. 4 (1964): 157–169.

Hebert, H. Josef, "Group: Cheney Task Force Eyed on Iraq Oil," Associated Press, July 18, 2003.

Hecht, Gabrielle, *The Radiance of France: Nuclear Power and National Identity after World War II* (Cambridge, MA: MIT Press, 2009 [1998]).

Heilprin, John, "White House Rejects Mandatory CO2 Caps," Associated Press, Feb. 2, 2007.

Heinberg, Richard, *The Party's Over: Oil, War, and the Fate of Industrial Societies*, 2nd ed. (Gabriola Island, BC: New Society Publishers, 2005).

Herbert, Bob, "A Price Too High?," *New York Times*, Mar. 19, 2011, A23.

Herrigel, Gary, *Industrial Constructions: The Sources of German Industrial Power* (New York: Cambridge University Press, 2000).

Herring, Horace, and Steve Sorrell, eds., *Energy Efficiency and Sustainable Consumption: The Rebound Effect* (New York: Palgrave Macmillan, 2009).

Hertog, Steffen, *Princes, Brokers, and Bureaucrats: Oil and the State in Saudi Arabia* (Ithaca, NY: Cornell University Press, 2010).

Hertsgaard, Mark, *Nuclear Inc.: The Men and Money behind Nuclear Energy* (New York: Pantheon, 1983).

Hewlett, Richard G., and Francis Duncan, *Atomic Shield, 1947/1952: A History of the United States Atomic Energy Commission*, vol. 2 (University Park: Pennsylvania State University Press, 1969).

Hewlett, Richard G., and Jack M. Holl, *Atoms and Peace and War 1953–1961: Eisenhower and the Atomic Energy Commission*, vol. 3 (Berkeley: University of California Press, 1989).

Hewlett, Richard G., and Oscar E. Anderson, Jr., *The New World, 1939/1946: A History of the United States Atomic Energy Commission*, vol. 1 (University Park: Pennsylvania State University Press, 1962).

Hise, Greg, *Magnetic Los Angeles: Planning the Twentieth-Century Metropolis* (Baltimore: Johns Hopkins University Press, 1997).

Hollis, Everett L., "The United States Atomic Energy Act of 1954—A Brief Survey," in *The Economics of Nuclear Power: Including Administration and Law*, J. Guéron, J. A. Lane, I. R. Maxwell, and J. R. Menke, eds. (New York: McGraw-Hill, 1957).

Holyoke, Thomas T., *Competitive Interests: Competition and Compromise in American Interest Group Politics* (Washington, D.C.: Georgetown University Press, 2011).

Hornby, Catherine, "Italy Plans to Reassess Nuclear Power in Few Years," Reuters, Apr. 26, 2011.

Hornstein, Jeffrey M., *A Nation of Realtors: A Cultural History of the Twentieth-Century American Middle Class* (Durham, NC: Duke University Press, 2005).

Hounshell, David A., *From the American System to Mass Production, 1800–1932: The Development of Manufacturing Technology in the United States* (Baltimore: Johns Hopkins University Press, 1984).

Houghton, John, *Global Warming: The Complete Briefing*, 4th ed. (New York: Cambridge University Press, 2009).

Humphries, Marc, ed., *U.S. Coal: A Primer on the Major Issues* (Hauppauge, NY: Novinka Books, 2004).

Hundley, Tom, "Iraq's Oil Industry Poised to Re-enter World Stage," *New York Times*, Feb. 15, 2010, Web.

Hunt, Michael H., *The American Ascendancy: How the United States Gained and Wielded Global Dominance* (Chapel Hill: University of North Carolina Press, 2007).

Hyman, Sidney, *Marringer S. Eccles: Private Entrepreneur and Public Servant* (Stanford, CA: Stanford University Graduate School of Business, 1976).

Ikenberry, John G., *Reasons of State: Oil Politics and the Capacities of American Government* (Ithaca, NY: Cornell University Press, 1988).

International Atomic Energy Agency, *Fast Breeder Reactors: Experience and Trends*, vol. 2 (Paris: International Atomic Energy Agency, 1986).

International Chamber of Commerce (ICC), *The International Chamber of Commerce* (Paris: International Chamber of Commerce, 2008).

International Chamber of Commerce (ICC) Commission on Energy and Environment, *Energy Efficiency with Case Studies* (Paris: International Chamber of Commerce, 2009).

International Energy Agency, *CO2 Emissions from Fuel Combustion*, 2010 ed. (Paris: International Energy Agency, 2010).

International Energy Agency, *Oil Market Report* (Paris: International Energy Agency, 2012 Mar. 14).

International Energy Agency, *World Energy Outlook 2009* (Paris: International Energy Agency, 2009).

International Energy Agency, *World Energy Outlook 2010* (Paris: International Energy Agency, 2010).

Jacobs, Andrew, "Chinese and British Officials Tangle in Testy Exchange over Climate Agreement," *New York Times*, Dec. 23, 2009, A10.

Jackson, Kenneth T., *Crabgrass Frontier: The Suburbanization of the United States* (New York: Oxford University Press, 1985).

Jolly, David, "Booming Car Sales in China May Bypass Diesels," *New York Times*, Oct. 1, 2010.

Jasper, James M., *Nuclear Politics: Energy and the State in the United States, Sweden, and France* (Princeton, NJ: Princeton University Press, 1990).

Jeffers, Thomas L., *Norman Podhoretz: A Biography* (New York: Cambridge University Press, 2010).

Johnson, Kirk, "A Battle over Uranium Bodes Ill for U.S. Debate," *New York Times*, Dec. 27, 2010, A1.

———, "Soaking Up the Sun to Squeeze Bills to Zero," *New York Times*, Feb. 15, 2011, D1.

Johnson, Peter J., and John Ensor Harr, *The Rockefeller Century: Three Generations of America's Greatest Family* (New York: Scribner, 1988).

Jones, Christopher F., "A Landscape of Energy Abundance: Anthracite Coal Canals and the Roots of American Fossil Fuel Dependence, 1820–1860," *Environmental History* 15, no. 3 (2010): 449–484.

Jones, Holway R., *John Muir and the Sierra Club: The Battle for Yosemite* (San Francisco: Sierra Club, 1965).

Joyner, Fred Bunyan, *David Ames Wells: Champion of Free Trade* (Cedar Rapids, IA: Torch Press, 1939).

Kahn, Joseph, "Cheney Promotes Increasing Supply as Energy Policy," *New York Times*, May 1, 2001, A1.

Kamber, Michael, and Taimoor Shag, "Iran Stops Fuel Delivery, Afghanistan Says, and Prices Are Rising," *New York Times*, Dec. 23, 2010, A12.

Kamieniecki, Sheldon, *Corporate America and Environmental Policy: How Often Does Business Get Its Way?* (Palo Alto, CA: Stanford University Press, 2006).

Kanter, James, "A Solar and Wind Revolution from a Land of Oil," *New York Times*, Mar. 13, 2011. Web.

———, "German Chancellor Calls for Tests of Europe's Nuclear Reactors," *New York Times*, Mar. 24, 2011, B3.

———, "Obstacles to Danish Wind Power," *International Herald Tribune*, Jan. 23, 2012, Finance sec., 18.

Karey, Gerald, "Ethanol Use Led to Higher Good Prices in US: CBO," *Platts Oilgram News*, Apr. 14, 2009, Markets & Data sec., p. 10.

Kay, Jane Holtz, *Asphalt Nation: How the Automobile Took Over America and How We Can Take It Back* (Berkeley: University of California Press, 1998).

Kenworthy, Jeffrey R., "Sustainable Urban Transport: Developing Sustainability Rankings and Clusters based on an International Comparison of Cities," in *Handbook of Sustainability Research*, vol. 20, Walter Leal Filho, ed. (New York: Peter Lang, 2005).

———, "Energy Use and CO2 Production in the Urban Passenger Transport Systems of 84 International Cities: Findings and Policy Implications," in *Urban Energy Transition from Fossil Fuels to Renewable Power*, Peter Droege, ed. (Amsterdam: Elsevier, 2008).

Kenworthy, Jeffrey R., and Felix B. Laube, with Peter Newman, Paul Barter, Tamim Raad, Chamlong Poboon, and Benedicto Guia, Jr., *An International Sourcebook of Automobile Dependence in Cities 1960–1990* (Boulder: University Press of Colorado, 1999).

Kirsch, David A., *The Electric Vehicle and the Burden of History* (New Brunswick, NJ: Rutgers University Press, 2000).

Klein, Maury, *The Genesis of Industrial America, 1870–1920* (New York: Cambridge University Press, 2007).

Knox, Paul L., *Metroburbia, USA* (Piscataway, NJ: Rutgers University Press, 2008).

Kolko, Gabriel, *The Triumph of Conservatism: A Reinterpretation of American History, 1900–1916* (New York: Free Press, 1977 [1963]).

Kramer, Andrew E., "Safety Issues Linger as Nuclear Reactors Shrink in Size," *New York Times*, Mar. 19, 2010, B1.

———, "Russia Is Seeking to Build Europe's Nuclear Plants," *New York Times*, Oct. 12, 2010, B11.

———, "Nuclear Industry in Russia Sells Safety, Taught by Chernobyl," *New York Times*, Mar. 23, 2011, B1.

———, "In Rebuilding Iraq's Oil Industry, U.S. Subcontractors Hold Sway," *New York Times*, June 17, 2011, B1.

———, "Here's an Easy $100 Billion Cut: Ending the Tax Subsidies for Oil and Ethanol Is Fiscally Sound and Right," *New York Times*, Aug. 8, 2011, A18.

Krasner, Stephen, *Defending the National Interest: Raw Materials Investments and U.S. Foreign Policy* (Princeton, NJ: Princeton University Press, 1978).

Krauss, Clifford, "Tapping a Trickle in West Texas," *New York Times*, Nov. 2, 2007, C1.

——, "Commodity Prices Tumble," *New York Times*, Oct. 14, 2008, B1.

——, "In Global Forecast, China Looms Large as Energy User and Maker of Green Power," *New York Times*, Nov. 10, 2010, B3.

——, "Why the Disruption of Libyan Oil Has Led to a Price Spike," *New York Times*, Feb. 24, 2011, B1.

——, "Ethanol Subsidies Besieged," *New York Times*, July 8, 2011, B1.

Krauss, Clifford, and Christine Hauser, "Oil Soars as Libyan Furor Shakes Markets," *New York Times*, Feb. 23, 2011, A1.

Krige, John, *American Hegemony and the Postwar Reconstruction of Science in Europe* (Cambridge, MA: MIT Press, 2006).

Kroenig, Matthew, *Exporting the Bomb: Technology Transfer and the Spread of Nuclear Weapons* (Ithaca, NY: Cornell University Press, 2010).

Krugman, Paul, "The Finite World," *New York Times*, Dec. 27, 2010, A19.

——, "Oligarchy, American Style," *New York Times*, Nov. 4, 2011, A31.

——, "Here Comes the Sun," *New York Times*, Nov. 7, 2011, A25.

Kryza, Frank T., *The Power of Light: The Epic Story of Man's Quest to Harness the Sun* (New York: McGraw-Hill, 2003).

Kurczy, Stephen, "International Energy Agency says 'Peak Oil' has Hit," *Christian Science Monitor*, Nov. 11, 2010.

Laird, Frank N., *Solar Energy, Technology Policy, and Institutional Values* (New York: Cambridge University Press, 2001).

Layzer, Judith A., "Deep Freeze: How Business Has Shaped the Global Warming Debate in Congress," in *Business and Environmental Policy*, Michael E. Kraft and Sheldon Kamieniecki, eds. (Cambridge, MA: MIT Press, 2007).

Leahy, Stephen, "Biofuels Boom Spurring Deforestation," *Inter Press Service*, Mar. 21, 2007.

Lester, Richard K. and David M. Hart, *Unlocking Energy Innovation: How America can Build a Low-Cost, Low Carbon Energy System* (Cambridge, MA: MIT Press, 2012).

Lichtblau, Eric, "Lobbyists' Long Effort to Revive Nuclear Industry Faces New Test," *New York Times*, Mar. 25, 2011, A1.

Lipton, Eric, and Clifford Krauss, "A U.S.-Backed Geothermal Plant in Nevada Struggles," *New York Times*, Oct. 3, 2011, B1.

——, "A Gold Rush of Subsidies in the Search for Clean Energy," *New York Times*, Nov. 12, 2011, A1.

Lisowski, Michael, "Playing the Two-Level Game: US President Bush's Decision to Repudiate the Kyoto Protocol," *Environmental Politics* 11, no. 4 (2002): 101–119.

Lochbaum, David, *The NRC and Nuclear Power Plant Safety in 2010: A Brighter Spotlight Needed* (Cambridge, MA: Union of Concerned Scientists, 2011).

Lodgaard, Sverre, *Nuclear Disarmament and Non-Proliferation: Towards a Nuclear-Weapon-Free World?* (New York: Routledge, 2010).

Logan, John R., and Harvey L. Molotch, *Urban Fortunes: The Political Economy of Place* (Berkeley: University of California Press, 1987 [2007]).

Lowi, Theodore J., *The End of Liberalism: The Second Republic of the United States* (New York: Norton, 1979).

Lucas, Nigel, *Western European Energy Policies: A Comparative Study of the Influence of Institutional Structures on Technical Change* (Oxford: Clarendon, 1985).

Luger, Stan, *Corporate Power, American Democracy, and the Automobile Industry* (New York: Cambridge University Press, 2000).

———, "Review of *Sloan Rules: Alfred P. Sloan and the Triumph of General Motors*," *American Historical Review* 110, no. 1 (2005): 174–175.

MacAvoy, Paul W., *The Natural Gas Market: Sixty Years of Regulation and Deregulation* (New Haven: Yale University Press, 2001).

Madrigal, Alexis, *Powering the Dream: The History and Promise of Green Technology* (Cambridge, MA: Da Capo Press, 2011).

Magat, Richard, *The Ford Foundation at Work: Philanthropic Choices, Methods, and Styles* (New York: Plenum, 1979).

Manley, John F., "Neo-pluralism: A Class Analysis of Pluralism I and Pluralism II," *American Political Science Review* 77, no. 2 (1983): 368–383.

Martinez, J. Michael, "The Carter Administration and the Evolution of American Nuclear Nonproliferation Policy, 1977–1981," *Journal of Policy History* 14, no. 3 (2002): 261–292.

Maslin, Mark, *Global Warming: A Very Short Introduction* (New York: Oxford University Press, 2009).

Mazzetti, Mark, "U.S. Intelligence Finding Says Iran Halted Its Nuclear Arms Effort in 2003," *New York Times*, Dec. 4, 2007, A1.

McConnell, Grant, *Private Power and American Democracy* (New York: Knopf, 1966).

McIntire, Mike, "Nonprofit Acts as a Stealth Business Lobbyist," *New York Times*, April 22, 2012, A1.

Metz, William D., and Allen L. Hammond, *Solar Energy in America* (Washington, DC: American Association for the Advancement of Science, 1978).

Miller, Arthur Selwyn, *The Modern Corporate State: Private Governments and the American Constitution* (Westport, CT: Greenwood, 1976).

Miller, Jerry, *Stockpile: The Story behind 10,000 Strategic Nuclear Weapons* (Annapolis, MD: Naval Institute Press, 2010).

Mills, Robin M., *Capturing Carbon: The New Weapon in the War Against Climate Change* (New York: Columbia University Press, 2011).

Mintz, Beth, and Michael Schwartz, *The Power Structure of American Business* (Chicago: University of Chicago Press, 1985).

Mitchell, William J., Christopher E. Borroni-Bird, and Lawrence D. Burns, *Reinventing the Automobile: Personal Urban Mobility for the 21st Century* (Cambridge, MA: MIT Press, 2010).

Mol, Arthur P. J., *Globalization and Environmental Reform: The Ecological Modernization of the Global Economy* (Cambridge, MA: MIT Press, 2001).

———, "Ecological Modernization and the Global Economy," *Global Environmental Politics* 2, no. 2 (2002): 92–115.

Mol, Arthur P. J., David A. Sonnenfeld, and Gert Spaargaren, eds., *The Ecological Modernisation Reader: Environmental Reform in Theory and Practice* (London: Routledge, 2009).

Molotch, Harvey, "The City as a Growth Machine: Towards of Political Economy of Place," *American Journal of Sociology* 82, no. 2 (1976): 309–322.

———, "Capital and Neighborhood in the United States," *Urban Affairs Quarterly* 14, no. 3 (1979): 289–312.

Moroney, John R., *Power Struggle: World Energy in the Twenty-First Century* (Westport, CT: Praeger, 2008).

Mouawad, Jad, "China's Growth Shifts the Geopolitics of Oil," *New York Times*, Mar. 19, 2010, B1.

———, "Natural Gas Now Viewed as Safer Bet," *New York Times*, Mar. 22, 2011, B1.

Mouawad, Jad, and Clifford Krauss, "Tremors From Libya Threaten to Rattle the Oil World," *New York Times*, Feb. 28, 2011, B1.

Mouawad, Jad, and Heather Timmons, "Trading Frenzy Adds to Jump In Price of Oil," New York Times, Apr. 29, 2006, A1.

Mueller, John, *Atomic Obsession: Nuclear Alarmism from Hiroshima to Al-Qaeda* (New York: Oxford University Press, 2010).

Muller, Peter O., *Contemporary Suburban America* (Englewood Cliffs, NJ: Prentice-Hall, 1981).

Myers, Steven Lee, "Tumult of Arab Spring Prompts Worries in Washington," *New York Times*, Sept. 18, 2011, A1.

National Energy Policy Development Group, *National Energy Policy* (Washington, DC: U.S. Government Printing Office, May 2001).

Nau, Henry, *National Politics and International Technology: Nuclear Reactor Development in Western Europe* (Baltimore: Johns Hopkins University Press, 1974).

Nayan, Rajiv, *The Nuclear Non-Proliferation Treaty and India* (New York: Routledge, 2011).

Nelkin, Dorothy, and Michael Pollak, *The Atom Besieged: Antinuclear Movements in France and Germany* (Cambridge, MA: MIT Press, 1981).

Nersesian, Roy L., *Energy for the 21st Century* (Armonk, NY: M.E. Sharpe, 2007).

"N.J. Nuclear Plant Closing Early," Associated Press, Dec. 8, 2010.

Newman, Peter, and Jeff Kenworthy, "Greening Urban Transportation," in *State of the World: Our Urban Future*, Linda Starke, ed. (New York: W.W. Norton, 2007).

Newman, Peter, Timothy Beatley, and Heather Boyer, *Resilient Cities: Responding to Peak Oil and Climate Change* (Washington, DC: Island Press, 2009).

Nexon, Daniel H., and Thomas Wright, "What's at Stake in the American Empire Debate," *American Political Science Review* 101, no. 2 (2007): 253–271.

Njølstad, Olav, *Nuclear Proliferation and International Order: Challenges to the Non-Proliferation Treaty* (New York: Routledge, 2010).

Nordhaus, William, *A Question of Balance* (Cambridge: MIT Press, 2008).

Nordlinger, Eric A., *On the Autonomy of the Democratic State* (Cambridge, MA: Harvard University Press, 1981).

Nye, David, *Image Worlds: Corporate Identities at General Electric, 1890-1930* (Cambridge, MA: MIT Press, 1985).

Olien, Diana Davids, and Roger M. Olien, *Oil in Texas: The Gusher Age, 1895–1945* (Austin: University of Texas Press, 2002).

Olien, Roger M., and Diana Davids Olien, *Oil and Ideology: The Cultural Creation of the American Petroleum Industry* (Chapel Hill: University of North Carolina Press, 2000).

Olney, Martha L., *Buy Now, Pay Later: Advertising, Credit, and Consumer Durables in the 1920s* (Chapel Hill: University of North Carolina Press, 1991).

Orsi, Richard J., "'Wilderness Saint' and 'Robber Baron': The Anomalous Partnership of John Muir and the Southern Pacific Company for Preservation of Yosemite National Park," *Pacific Historian* 29 (summer–fall 1985): 136–152.

Painter, David S., *Oil and the American Century: The Political Economy of U.S. Foreign Oil Policy, 1941–1954* (Baltimore: Johns Hopkins University Press, 1986).

Palmer, Steven, *Launching Global Health: The Caribbean Odyssey of the Rockefeller Foundation* (Ann Arbor: University of Michigan Press, 2010).

Palz, Wolfgang, ed., *Power for the World: The Emergence of Electricity from the Sun* (Singapore: Pan Stanford Publishing, 2011).

Panel on the Impact of the Peaceful Uses of Atomic Energy, *Peaceful Uses of Atomic Energy*, vols. 1–2 (Washington, DC: Government Printing Office, 1956).

Parmar, Inderjeet, "American Foundations and the Development of International Knowledge Networks," *Global Networks* 2, no. 1 (2002): 13–30.

Parra, Francisco, *Oil Politics: A Modern History of Petroleum* (New York: I.B. Tauris, 2004).

Paterson, Matthew, *Automobile Politics* (New York: Cambridge University Press, 2007).

Pauly, Jr., Robert J., *U.S. Foreign Policy and the Persian Gulf: Safeguarding American Interests through Selective Multilateralism* (Burlington, VT: Ashgate, 2005).

Paxson, Frederic L., "The American Highway Movement, 1916–1935," *American Historical Review* 51, no. 2 (1946): 239–241.

Pfau, Richard, *No Sacrifice Too Great: The Life of Lewis L. Strauss* (Charlottesville: University Press of Virginia, 1984).

Philip, George, *The Political Economy of International Oil* (Edinburgh: Edinburgh University Press, 1994).

Pilat, Joseph F., ed., *Atoms for Peace: A Future after Fifty Years* (Baltimore: Johns Hopkins University Press, 2007).

Podobnik, Bruce, *Global Energy Shifts: Fostering Sustainability in a Turbulent Age* (Philadelphia: Temple University Press, 2006).

Polmar, Norman, *U.S. Nuclear Arsenal: A History of Weapons and Delivery Systems since 1945* (Annapolis, MD: Naval Institute Press, 2009).

Poulantzas, Nico, *Political Power and Social Classes* (London: New Left Books, 1973).

Powell, James Lawrence, *The Inquisition of Climate Science* (New York: Columbia University Press, 2011).

Power, Max S., *America's Nuclear Wastelands: Politics, Accountability, and Cleanup* (Pullman: Washington State University Press, 2008).

Proceedings of the World Symposium on Applied Solar Energy, Phoenix, AZ, Nov. 1–5, 1955 (San Francisco: Jorgenson & Co., 1956).

Price, Terrence, *Political Electricity: What Future for Nuclear Energy?* (New York: Oxford University Press, 1990).

Radford, Gail, *Modern Housing for America: Policy Struggles in the New Deal Era* (Chicago: University of Chicago Press, 1996).

Rajan, Sudhir, *The Enigma of Automobility: Democratic Politics and Pollution Control* (Pittsburgh: University of Pittsburgh Press, 1996).

Report of the Committee on Recent Economic Changes, of the President's Conference on Unemployment, *Recent Economic Changes in the United States*, vols. 1–2 (New York: McGraw-Hill, 1929).

Report of the President's Conference on Unemployment (Washington, DC: Government Printing Office, 1921).

Rich, Andrew, *Think Tanks, Public Policy, and the Politics of Expertise* (New York: Cambridge University Press, 2004).

Righter, Robert, *Windfall: Wind Energy in America Today* (Norman: University of Oklahoma Press, 2011).

Roberts, Paul, *The End of Oil: On the Edge of a Perilous New World* (New York: Houghton Mifflin, 2004).

———, afterward to Paul Roberts, *The End of Oil: On the Edge of a Perilous New World* (New York: Houghton Mifflin, 2005).

Rockefeller Foundation Annual Report, 1956 (New York: Rockefeller Foundation, 1957).

Roelofs, Joan, *Foundations and Public Policy: The Mask of Pluralism* (Albany: State University of New York Press, 2003).

Rome, Adam, *The Bulldozer in the Countryside: Suburban Sprawl and the Rise of American Environmentalism* (Cambridge: Cambridge University Press, 2001).

Romero, Simon, "Oil-Rich Norwegians Take World's Highest Gasoline Prices in Stride," *New York Times*, Apr. 30, 2005, C1.

———, "New Fields May Propel Americas to Top of Oil Companies' Lists," *New York Times*, Sept. 20, 2011, A1.

Romm, Joseph J., *The Hype about Hydrogen: Fact and Fiction in the Race to Save the Climate* (Washington, DC: Island Press, 2004).

Rosen, Elliot, *Roosevelt, the Great Depression, and the Economics of Recovery* (Charlottesville: University of Virginia Press, 2005).

Rosenthal, Elisabeth, "Studies Call Biofuels a Greenhouse Threat," *New York Times*, Feb. 8, 2008, A9.

———, "African Huts Far from the Grid Glow with Renewable Power," *New York Times*, Dec. 25, 2010, A1.

———, "Rush to Use Crops as Fuel Raises Food Prices and Hunger Fears," *New York Times*, Apr. 7, 2011, A1.

———, "Across Europe, Irking Drivers Is Urban Policy," *New York Times*, June 27, 2011, A1.

———, "Germany Dims Nuclear Plants, but Hopes to Keep Lights On," *New York Times*, Aug. 30, 2011, A1.

Rutledge, Ian, *Addicted to Oil: America's Relentless Drive for Energy Security* (New York: IB Tauris, 2005).

Sanders, M. Elizabeth, *The Regulation of Natural Gas: Policy and Politics, 1938–1978* (Philadelphia: Temple University Press, 1981).

Sanger, David E., "U.S. and Allies Plan More Sanctions Against Iran," *New York Times*, Dec. 11, 2010, A6.

———, "Iran Moves to Shelter Its Nuclear Fuel Program," *New York Times*, Sept. 2, 2011, A4.

Sanger, David E., and Matthew L. Wald, "Radioactive Releases in Japan Could Last Months, Experts Say," *New York Times*, Mar. 14, 2011, A1.

Sanger, David E., and William J. Broad, "Iran Says It Will Speed Up Uranium Enrichment," *New York Times*, June 9, 2011, A14.

Sanger, David E., and William J. Broad, "Survivor of Attack Leads Nuclear Effort in Iran," *New York Times*, July 23, 2011, A4.

Schaller, Michael, *Altered States: The United States and Japan since the Occupation* (New York: Oxford University Press, 1997).

Schmidheiny, Stephen, and Federico Zorraquin, with the World Business Council for Sustainable Development, *Financing Change: The Financial Community, Ecoefficiency, and Sustainable Development* (Cambridge, MA: MIT Press, 1996).

Schneider, Keith, "Midwest Emerges as Center for Clean Energy," *New York Times*, Dec. 1, 2010, B8.

Schrepfer, Susan R., *The Fight to Save the Redwoods: A History of Environmental Reform, 1917–1978* (Madison: University of Wisconsin Press, 1983).

Semple, Jr., Robert B., "Oil and Gas Had Help. Why Not Renewables?" *New York Times*, Oct. 16, 2011, SR10.

Seyoum, Belay, *Export-Import Theory, Practices, and Procedures*, 2nd ed. (New York: Routledge, 2008).

Shaffer, Brenda, *Energy Politics* (Philadelphia: University of Pennsylvania Press, 2009).

Shaffer, Ed, *The United States and the Control of World Oil* (New York: St. Martin's Press, 1983).

Sheehan, Molly O'Meara, *City Limits: Putting the Brakes on Sprawl* (Washington, DC: Worldwatch Institute, 2001).

Shor, Francis, *Dying Empire: U.S. Imperialism and Global Resistance* (New York: Routledge, 2010).

Shulman, Seth, *Undermining Science: Suppression and Distortion in the Bush Administration* (Berkeley: University of California Press, 2006).

Siebert, Horst, *The German Economy: Beyond the Social Market* (Princeton, NJ: Princeton University Press, 2005).

Siegelbaum, Lewis H., *Cars for Comrades: The Life of the Soviet Automobile* (Ithaca, NY: Cornell University Press, 2008).

Simmons, Matthew R., *Twilight in the Desert: The Coming Saudi Oil Shock and the World Economy* (New York: Wiley, 2005).

Skocpol, Theda, *States and Social Revolutions* (Cambridge: Cambridge University Press, 1979).

———, "Bringing the State Back In: Strategies of Analysis in Current Research," in *Bringing the State Back In*, Peter Evans, Dietrich Rueschemeyer, and Theda Skocpol, eds. (Cambridge: Cambridge University Press, 1985).

———, "A Brief Response [to G. William Domhoff]," *Politics and Society* 15, no. 3 (1986/87): 331–332.

———, *Protecting Soldiers and Mothers: The Political Origins of Social Policy in the United States* (Cambridge: Harvard University Press, 1992).

Skocpol, Theda, Marshall Ganz, and Ziad Munson, "A Nation of Organizers: The Institutional Origins of Civic Voluntarism in the United States," *American Political Science Review* 94, no. 3 (2000): 527–546.

Skowronek, Stephen, *Building a New American State: The Expansion of National Administrative Capacities, 1877–1920* (Cambridge: Cambridge University Press, 1982).

Smil, Vaclav, *Energy Myths and Realities: Bringing Science to the Energy Policy Debate* (Washington, DC: AEI Press, 2010).

———, *Energy Transitions: History, Requirements, Prospects* (Denver, CO: Praeger, 2010).

Snell, Bradford, *American Ground Transport* (Washington, DC: U.S. Government Printing Office, 1974).

St. Clair, David J., *The Motorization of American Cities* (New York: Praeger, 1986).

Stilgoe, John, *Borderland: Origins of the American Suburb, 1820–1939* (New Haven, CT: Yale University Press, 1988).

Steele, Brian C. H., and Angelika Heinzel, "Materials for Fuel-Cell Technologies," *Nature* 414 (Nov. 2001): 345–352.

Stewart, Richard Burleson, and Jane Bloom Stewart, *Fuel Cycle to Nowhere: U.S. Law and Policy on Nuclear Waste* (Nashville, TN: Vanderbilt University Press, 2011).

Stoett, Peter, "Toward Renewed Legitimacy? Nuclear Power, Global Warming, and Security," *Global Environmental Politics* 3, no. 1 (2003): 99–116.

Stokes, Doug, and Sam Raphael, *Global Energy Security and American Hegemony* (Baltimore: Johns Hopkins University Press, 2010).

Strum, Harvey, "Eisenhower's Solar Energy Policy," *The Public Historian* 6, no. 2 (1984): 37–50.

———, "The Association for Applied Solar Energy/Solar Energy Society, 1954–1970," *Technology and Culture* 26, no. 3 (1985): 571–578.

Strum, Harvey, and Fred Strum, "American Solar Energy Policy, 1952–1982," *Environmental Review* 7 (summer 1983): 135–153.

Swenson-Wright, John, *Unequal Allies? United States Security and Alliance Policy toward Japan, 1945-1960* (Stanford, CA: Stanford University Press, 2005).

Sylves, Richard T., *The Nuclear Oracles: A Political History of the General Advisory Committee of the Atomic Energy Commission, 1947–1977* (Ames: Iowa State University Press, 1987).

Tabuchi, Hiroko, "Japanese Workers Braved Radiation for a Temp Job," *New York Times*, Apr. 10, 2011, A1.

———, "Japan Premier Wants Shift Away from Nuclear Power," *New York Times*, July 14, 2011, A6.

————, "Japan Passes Law Supporting Stricken Nuclear Plant's Operator," *New York Times*, Aug. 4, 2011, A8.

Tabuchi, Hiroko, and Keith Bradsher, "Japan Nuclear Disaster Put on Par With Chernobyl," *New York Times*, Apr. 12, 2011, A10.

Tabuchi, Hiroko, and Matthew L. Wald, "Partial Meltdowns Presumed at Crippled Reactors," *New York Times*, Mar. 14, 2011, A7.

Tatum, Jesse S., *Energy Possibilities: Rethinking Alternatives and the Choice-Making Process* (Albany: State University of New York Press, 1995).

"The End of Clean Energy Subsidies?" *New York Times*, May 6, 2012, SR12.

Thompson, C. Bradley, with Yaron Brook, *Neoconservatism: An Obituary for an Idea* (Boulder, CO: Paradigm Publishers, 2010).

Timmons, Heather, and Vikas Bajaj, "Emerging Economies Move Ahead with Nuclear Plans," *New York Times*, Mar. 15, 2011, B1.

Todd, Emmanuel, *After the Empire: The Breakdown of the American Order*, trans. C. Jon Delogu (New York: Columbia University Press, 2003).

Toke, David, *Ecological Modernisation and Renewable Energy* (New York: Palgrave Macmillan, 2011).

Touraine, Alain, *Anti-nuclear Protest: The Opposition to Nuclear Energy in France* (New York: Cambridge University Press, 1983).

Truman, David B., *The Governmental Process: Political Interests and Public Opinion* (New York: Knopf, 1951).

Twentieth Century Fund Task Force on the International Oil Crisis, *Paying for Energy* (New York: McGraw-Hill, 1975).

Twentieth Century Fund Task Force on United States Energy Policy, *Providing for Energy* (New York: McGraw-Hill, 1977).

Uchitelle, Louis, "Goodbye, Production (and Maybe Innovation)," *New York Times*, Dec. 24, 2006, sec. 3, p. 4.

Urbina, Ian, "Regulators Seek Records on Claims for Gas Wells," *New York Times*, July 30, 2011, A13.

————, "New Report by Agency Lowers Estimates of Natural Gas in U.S.," *New York Times*, Jan. 29, 2012, A16.

Useem, Michael, *The Inner Circle: Large Corporations and the Rise of Business Political Activity in the U.S. and U.K.* (Oxford: Oxford University Press, 1984).

Vaïsse, Justin, *Neoconservatism: The Biography of a Movement* (Cambridge, MA: Harvard University Press, 2010).

Van Til, Jon, *Living with Energy Shortfall* (Boulder, CO: Westview, 1982).

Vandenbosch, Robert, and Susanne E. Vandenbosch, *Nuclear Waste Stalemate: Political and Scientific Controversies* (Salt Lake City: University of Utah Press, 2007).

Vasi, Ion Bogdan, *Winds of Change: The Environmental Movement and the Global Development of the Wind Energy Industry* (New York: Oxford University Press, 2011).

Victor, David G., Amy M. Jaffe, and Mark H. Hayes, eds., *Natural Gas and Geopolitics: From 1970 to 2040* (New York: Cambridge University Press, 2006).

Vietor, Richard H., *Environmental Politics and the Coal Coalition* (College Station: Texas A&M University Press, 1980).

———, *Engery Policy in America since 1945* (New York: Cambridge University Press, 1984).

Wald, Matthew L., "Will Hydrogen Clear the Air? Maybe Not, Say Some," *New York Times*, Nov. 12, 2003, C1.

———, "When It Comes to Replacing Oil Imports, Nuclear Is No Easy Option, Experts Say," *New York Times*, May 9, 2005, A14.

———, "As Nuclear Waste Languishes, Expense to U.S. Rises," *New York Times*, Feb. 17, 2008, A22.

———, "U.S. Rejects Nuclear Plant over Design of Key Piece," *New York Times*, Oct. 16, 2009, A13.

———, "Fee Dispute Hinders Plan for Reactor," *New York Times*, Oct. 10, 2010, A21.

———, "Sluggish Economy Curtails Prospects for Building Nuclear Reactors," *New York Times*, Oct. 11, 2010, B1.

———, "Vermont Nuclear Plant Up for Sale," *New York Times*, Nov. 5, 2010, B9.

———, "Study of Baby Teeth Sees Radiation Effects," *New York Times*, Dec. 14, 2010, D2.

———, "New Interest in Turning Gas to Diesel," *New York Times*, Dec. 24, 2010, B1.

———, "Administration to Push for Small 'Modular' Reactors," *New York Times*, Feb. 13, 2011, A29.

———, "3 States Challenge Federal Policy on Storing Nuclear Waste," *New York Times*, Feb. 16, 2011, A23.

———, "Showdown on Vermont Nuclear Plant's Fate," *New York Times*, Mar. 11, 2011, A19.

———, "Japan Orders Evacuation near 2nd Nuclear Plant," *New York Times*, Mar. 12, 2011, A10.

———, "Report Urges Storing Spent Nuclear Fuel, Not Reprocessing It," *New York Times*, Apr. 26, 2011, A16.

———, "Court Won't Intervene in Fate of Nuclear Dump," *New York Times*, July 2, 2011, A13.

———, "A Safer Nuclear Crypt," *New York Times*, July 6, 2011, B1.

———, "U.S. Backs Project to Produce Fuel from Corn Waste," *New York Times*, July 7, 2011, B10.

———, "N.R.C. Lowers Estimate of How Many Would Die in Meltdown," *New York Times*, July 30, 2011, A14.

———, "Alabama Nuclear Reactor, Partly Built, to Be Finished," *New York Times*, Aug. 19, 2011, A12.

———, "U.S. Backs New Loans for Projects on Energy," *New York Times*, Sept. 29, 2011, A14.

———, "Batteries at a Wind Farm Help Control Output," *New York Times*, Oct. 29, 2011, B3.

———, "Taming Unruly Wind Power," *New York Times*, Nov. 5, 2011, B1.

———, "Loan Request by Uranium-Enrichment Firm Upends Politics as Usual," *New York Times*, Nov. 25, 2011, B5.

———, "Storehouses for Solar Energy Can Step In When the Sun Goes Down," *New York Times*, Jan. 3, 2012, B1.

———, "A Fine for Not Using a Biofuel That Doesn't Exist," *New York Times*, Jan. 10, 2012, B1.

Walker, Samuel, *The Road to Yucca Mountain: The Development of Radioactive Waste Policy in the United States* (Los Angeles: University of California Press, 2009).

Weale, Albert, *The New Politics of Pollution* (New York: Manchester University Press, 1992).

Weiss, Marc, *The Rise of the Community Builders: The American Real Estate Industry and Urban Land Planning* (New York: Columbia University Press, 1987).

Weinstein, James, *The Corporate Ideal in the Liberal State: 1900–1918* (Boston: Beacon Press, 1968).

Wells, David A., *Practical Economics: A Collection of Essays Respecting Certain of the Recent Economic Experiences of the United States* (New York: G.P. Putnam's Sons, 1885).

Wetherly, Paul, *Marxism and the State: An Analytical Approach* (New York: Palgrave, 2005).

Wetherly, Paul, Clyde W. Barrow, and Peter Burnham, eds., *Class, Power and the State in Capitalist Society: Essays on Ralph Miliband* (New York: Palgrave MacMillan, 2008).

Williams, William Appleman, *The Roots of the Modern American Empire* (New York: Random House, 1969).

Wills, John, *Conservation Fallout: Nuclear Protest at Diablo Canyon* (Reno: University of Nevada Press, 2006).

Wingrove, Josh, "Can Coal Come Clean or Is Wind the Future?" *Globe and Mail*, Sept. 13, 2011, A3.

Wilson, Joan Hoff, *American Business & Foreign Policy, 1920-1933* (Lexington: University Press of Kentucky, 1971).

Winks, Robin W., *Laurance S. Rockefeller: Catalyst for Conservation* (Washington, DC: Island Press, 1997).

Winters, Jeffrey A., and Benjamin I. Page, "Oligarchy in the United States," *Perspectives on Politics* 7, no. 4 (2009): 731–751.

Wood, Ellen Meiksins, *The Origin of Capitalism* (New York: Monthly Review Press, 1999).

———, *Empire of Capital* (New York: Verso, 2003).

Woody, Todd, "Solar Power Projects Face Potential Hurdles," *New York Times*, Oct. 29, 2010, B1.

———, "G.E. Plans to Build Largest Solar Panel Plant in U.S.," *New York Times*, Apr. 7, 2011, B3.

World Business Council for Sustainable Development, *Mobility 2030: Meeting the Challenge of Sustainability* (report overview) (Washington, DC: World Business Council for Sustainable Development, 2004).

———, *Pathways to 2050: Energy and Climate Change* (Washington, DC: World Business Council for Sustainable Development, 2005).

———, *Policy Directions to 2050: Energy & Climate* (Washington, DC: World Business Council for Sustainable Development, 2007).

———, *Vision 2050: The New Agenda for Business* (Washington, DC: World Business Council for Sustainable Development, 2010).

"World Oil Demand to Peak before Supply—BP," Reuters, Jan. 16, 2008.

Worth, Robert F., "Unrest Encircles Saudis, Stoking Sense of Unease," *New York Times*, Feb. 20, 2011, A13.

Yago, Glen, *The Decline of Transit: Urban Transportation in German and U.S. Cities, 1900–1970* (New York: Cambridge University Press, 1984).

Yergin, Daniel, *The Prize: The Epic Quest for Oil, Money, and Power* (New York: Simon & Schuster, 1991).

———, *The Quest: Energy, Security, and the Remaking of the Modern World* (New York: Penguin, 2011).

Yetiv, Steve A., *Crude Awakenings: Global Oil Security and American Foreign Policy* (Ithaca, NY: Cornell University Press, 2004).

———, *Explaining Foreign Policy: U.S. Decision-Making in the Gulf Wars* (Baltimore: Johns Hopkins University Press, 2011).

Yong, William, "Gas Prices Soar in Iran as Subsidy Is Reduced," *New York Times*, Dec. 20, 2010, A6.

York, Richard, and Eugene A. Rosa, "Key Challenges to Ecological Modernization Theory," *Organization & Environment* 16, no. 3 (2003): 273–288.

Zachary, G. Pascal, *Endless Frontier: Vannevar Bush: Engineer of the American Century* (New York: Free Press, 1997).

Zaun, Todd, "Honda Tries to Spruce Up a Stodgy Image," *New York Times*, Mar. 19, 2005, C3.

Zedalis, Rex J., *The Legal Dimensions of Oil and Gas in Iraq: Current Reality and Future Prospects* (New York: Cambridge University Press, 2009).

Zeller, Tom, Jr., "U.S. Nuclear Plants Have Same Risks, and Backups, as Japan Counterparts," *New York Times*, Mar. 14, 2011, A10.

———, "With U.S. Nuclear Plants under Scrutiny, Too, a Report Raises Safety Concerns," *New York Times*, Mar. 18, 2011, A11.

Zimmerman, Julian H., *The FHA Story in Summary, 1934–1959* (Washington, DC: U.S. Federal Housing Administration, 1959).

Index